STP 993

Mechanical Relaxation of Residual Stresses

Leonard Mordfin, editor

ASTM
1916 Race Street
Philadelphia, PA 19103

Library of Congress Cataloging-in-Publication Data

Mechanical relaxation of residual stresses / Leonard Mordfin, editor.
(STP; 993)
Papers from the International Symposium on Mechanical Relaxation of Residual Stresses, held in Cincinnati, Ohio, Apr. 30, 1987 and sponsored by ASTM Committee E-28 on Mechanical Testing.
Includes bibliographies and index.
"ASTM publication code number (PCN) 04-993000-23."
ISBN 0-8031-1166-5
1. Residual stresses—Congresses. 2. Stress relaxation--Congresses. I. Mordfin, Leonard. II. International Symposium on Mechanical Relaxation of Residual Stresses (1987: Cincinnati, Ohio) III. American Society for Testing and Materials. Committee E-28 on Mechanical Testing. IV. Series: ASTM special technical publication ; 993.
TA417.6.M426 1988 88-15450
620.1′124—dc 19. CIP

NOTE

The Society is not responsible, as a body, for the statements and opinions advanced in this publication

Peer Review Policy

Each paper published in this volume was evaluated by three peer reviewers. The authors addressed all of the reviewers' comments to the satisfaction of both the technical editor(s) and the ASTM Committee on Publications.

The quality of the papers in this publication reflects not only the obvious efforts of the authors and the technical editor(s), but also the work of these peer reviewers. The ASTM Committee on Publications acknowledges with appreciation their dedication and contribution of time and effort on behalf of ASTM.

Printed in Baltimore, MD
July 1988

Foreword

The International Symposium on Mechanical Relaxation of Residual Stresses was presented at Cincinnati, Ohio, 30 April 1987. It was sponsored by ASTM Committee E 28 on Mechanical Testing. Leonard Mordfin, National Bureau of Standards, Gaithersburg, Maryland, served as chairman of the symposium and as the editor of this publication.

Contents

Introduction

The International Symposium on Mechanical Relaxation of Residual Stresses was convened on April 30, 1987, in Cincinnati, Ohio, with speakers from five countries participating. The objective was to obtain a better understanding of the processes by which residual stresses are relaxed by certain mechanical treatments. This volume presents the peer reviewed and edited proceedings of that specialists' symposium.

One of the interesting adjectives that is often used to describe residual stresses is "insidious." This is because residual stresses are present in virtually every solid material or component but, because we can't see them and because without some careful and complex measurements we don't know how severe they might be, we sometimes tend to forget that they are there. Yet, their effects can be significant, even catastrophic.

In spite of their insidious nature, we do know some basic and important things about residual stresses. We know, for example, that residual stresses and applied stresses are algebraically additive within the elastic range and, thus, residual stresses can be beneficial as well as detrimental. Fatigue damage, crack propagation, and stress corrosion are tensile phenomena and, therefore, tensile residual stresses may contribute to the development of these failure modes. Conversely, compressive residual stresses are beneficial in that they tend to inhibit these occurrences. However, large residual stresses, whether they are tensile or compressive, can cause dimensional instability, either through creep or by their redistribution as a result of machining.

It is, of course, desirable to reduce residual stresses that are detrimental. One means for doing this, which is usually effective and economical, is thermal stress relief. Sometimes, however, this method is not practical because a thermal treatment would be detrimental to other characteristics of the object, or because the object is too large, or for any of a variety of other reasons. In such cases, mechanical stress relaxation is often a viable alternative. Cold working and vibration are two common means for achieving mechanical stress relaxation. When the residual stress distribution in a given object is beneficial, then it is likely that no intentional effort would be made to relieve these stresses. However, depending upon the operational stresses imposed upon the object while it is in service, mechanical stress relaxation could take place anyway, say, by the effects of cyclic loads.

Whether mechanical stress relaxation takes place intentionally or unintentionally it would be very helpful in many applications to understand how it happens and what its magnitude is. In general, this kind of information is sparse. Available knowledge about mechanical stress relaxation is largely qualitative. Very few studies of the mechanical relaxation of residual stresses have involved actual measurements of the reduction in residual stresses as a result of mechanical treatments. On the contrary, most of what is known or believed to be true about mechanical stress relaxation has been inferred from indirect observations of the effects of residual stresses. In other words, changes in fatigue, fracture, or stress corrosion behavior, or in dimensional stability, which follow mechanical treatments of the kinds mentioned, are simply attributed to changes in the residual stresses. Clearly, this kind of knowledge is neither complete nor reliable. Without an understanding of the mechanics of residual stress relaxation and a quantitative grasp of its magnitudes, the designer is unable to rely upon the benefits of beneficial residual stresses nor can he avoid overdesigning to allay the possible detriments of detrimental residual stresses. Similarly, the maintenance engineer and the in-service inspector can never be certain under these circumstances about the continued integrity of a structure or component.

This inadequate understanding of mechanical stress relaxation has developed from the difficulties in measuring residual stresses accurately and reliably. Substantial progress in recent years has mitigated some of these difficulties. Today it is frequently possible to measure residual stresses rapidly and economically as well as accurately and reliably. New standard test methods as well as technological developments have effected this advancement in measurement capabilities. ASTM Standard Method for Determining Residual Stresses by the Hole-Drilling Strain-Gage Method (E 837-85) and ASTM Standard Method for Verifying the Alignment of X-Ray Diffraction Instrumentation for Residual Stress Measurement, (E 915-85) are noteworthy in this context.

With these new capabilities, Subcommittees E28.11 and E28.13 in ASTM Committee E-28 on Mechanical Testing decided about two years ago that it was now feasible to seek a more definitive understanding of the mechanical relaxation of residual stresses, and plans for an international specialists' symposium on this specific, important, topic were initiated.

Through the efforts of Matt Lieff, then ASTM staff manager for Committee E-28, and his counterpart in ASM International, agreements were reached to hold the symposium immediately following the ASM Conference on Residual Stress in Design, Process and Material Selection. The ASM conference was scheduled for April 27–29 and the ASTM symposium for April 30, 1987, in Cincinnati, Ohio. In view of the relatively general nature of the ASM conference and the narrowly focused scope of the ASTM symposium, no unpleasant competition for papers was anticipated and none developed. On the contrary, the back-to-back timing probably benefitted both meetings by attracting additional attendees to the unusual double feature.

No papers were *invited* for the ASTM symposium since, frankly, the members of the organizing committee were not aware of any recent or on-going research activities that involved actual measurements of residual stress relaxation. Total reliance was placed on an international call for *contributed* papers. Nevertheless, the committee was adamant in its intention to accept only papers that specifically included measurements of the relaxation of residual stresses due to mechanical treatments, or theoretical analyses of this phenomenon. Needless to say, it is always difficult to reject good papers simply because they do not address the intended theme of a symposium or a conference, and that is why it is common to see irrelevant papers on symposium and conference programs. The rejection task was made relatively painless in the case of this symposium because of a fortunate set of circumstances. The chairman of the symposium was simultaneously serving on the organizing committee for the ASM conference and was also arranging a session on residual stress for the 6th International Conference on Pressure Vessel Technology (ICPVT-6). Thus, good papers which were submitted to the ASTM symposium but did not adequately address the very specific theme of that symposium were diverted to one of the other conferences, which had comparatively broad themes. As a result, out of sixteen papers submitted to the ASTM symposium, seven were accepted for presentation and subsequent publication. Three others were accepted for the ASM conference and one for the ICPVT-6.

Although only seven papers were considered acceptable for the symposium, they represented a high-quality, balanced mix. Three of the papers addressed the relaxation of residual stress by the application of cold working, two papers dealt with vibratory stress relief, and two with the relaxation of residual stresses that accompanies cyclic loading. All of these papers are presented in this volume, together with the abstract of a paper which was only presented orally.

To the editor's knowledge, this is the only book specifically devoted to the mechanics of the mechanical relaxation of residual stresses. There is reason to believe that this volume will serve to eliminate many of the misconceptions that have existed regarding the mechanical relaxation of residual stresses and will also help to stimulate its use in applications where such treatments would be desirable. The development of standard test methods for evaluating mechanical stress relaxation may also be feasible now.

It is a pleasure to acknowledge the diligent efforts of the authors of the papers—their cooperation, despite separation by oceans, has been exceptional. Thanks are also due to all of the other people who helped produce this book; the hand-picked, expert reviewers whose evaluations and critical comments immeasurably enhanced the final versions of the papers, and the ASTM editorial staff upon whom we have all come to rely so heavily.

Leonard Mordfin
United States Department of Commerce
National Bureau of Standards
Gaithersburg, Maryland; symposium chairman and editor.

Stress Relaxation by Cold Working

W. E. Nickola[1]

Residual Stress Alterations via Cold Rolling and Stretching of an Aluminum Alloy

REFERENCE: Nickola, W. E., "**Residual Stress Alterations via Cold Rolling and Stretching of an Aluminum Alloy,**" *Mechanical Relaxation of Residual Stresses, ASTM STP 993,* L. Mordfin, Ed., American Society for Testing and Materials, Philadelphia, 1988, pp. 7–18.

ABSTRACT: A solution heat-treated, cold-water quenched experimental aluminum alloy was subjected to two sequential mechanical treatments: an 11.5 % reduction by cold rolling followed by a 1.25 % cold stretching. The layer-removal and hole-drilling strain-gage methods of residual stress analysis were used, independently, to quantify both the changes in residual stresses produced by cold rolling and the reduction of stress produced by cold stretching. The initial residual stress condition, surface compression with mid-plane tension, was reversed by the cold rolling mechanical treatment to a condition of tensile surface stresses with mid-plane compression. Cold stretching produced a marked reduction in the magnitudes of the cold-rolling-induced residual stress. The independent experimental results of the layer-removal and hole-drilling strain-gage methods corroborated the effectiveness of stretching as a method for reducing residual stresses.

KEY WORDS: residual stress, mechanical relaxation, cold rolling, stretch, mechanical treatments, layer-removal, hole-drilling, stress reduction

Nomenclature

A, B	Coefficients defined in Eq 3
$\bar{a}, \bar{b}$	Nondimensional coefficients per Eq 5, from Schajer [9]
D_o	Diameter of drilled hole
z	Depth of drilled hole
α	Direction angle relative to σ_x (Fig. 2)
β	Principal stress direction as defined in Fig. 3 and Eq 4
ϵ_r	Radial strain at point P of Fig. 2
$\epsilon_1, \epsilon_2, \epsilon_3$	Measured relieved strains from residual stress strain gage rosettes
σ_x, σ_y	Biaxial principal stresses of Fig. 2
σ_1, σ_2	Maximum and minimum principal stresses, respectively
ν, E	Poisson's ratio and elastic modulus, respectively
ϕ_x, ϕ_y	Measured curvatures

Introduction

General theoretical methods for analysis of residual stresses are not, at present, practical engineering design tools; however, measurement methods are used successfully to establish a quantitative understanding of stress magnitude. Residual-stress measurements in an industrial setting are most frequently accomplished using mechanical or X-ray diffraction

[1] Measurements Group, Inc., Raleigh, North Carolina 27611.

procedures. There are many variations of the mechanical approach and while some are quite old [*1,2,3*], others have been introduced more recently. However, all mechanical approaches involve the removal of material, at least to some degree, and the taking of strain or deformation measurements is a normal requirement.

These different experimental methods for measuring residual stress are valuable engineering tools that can be used to establish a technical understanding of residual stress gradients and their magnitudes. They are highly suitable for quantitative evaluation of the various mechanical treatments, such as mechanical vibration and ultrasonic excitation, that are presently used to reduce undesirable residual stresses. In many mechanical treatment applications, the reduction of residual stresses is inferred from observed improvement in product performance. But while improved performance is certainly the objective, residual stress measurements offer the additional advantage of providing a quantitative appreciation that can often be related to the degree, or amount, of mechanical treatment that is required.

This technical paper chronicles the determination and subsequent reduction of residual stresses in an aluminum alloy plate with the initial condition existing after solution heat treatment with a cold water quench, followed by a room temperature age. There were then two sequential mechanical treatments: 11.5 % reduction by cold rolling, and finally a 1.25 % cold stretch. Employing the layer-removal method [*4*] and the hole-drilling strain-gage method [*4,5*] residual stresses were measured both before and after the mechanical treatments.

The Experimental Methods

Layer Removal Method

Treuting and Reed [*4*] derived and documented a general method of layer removal that is applicable to homogeneous elastic materials; a brief summary of this method is included here.

A plate with maximum and minimum principal residual stresses (σ_x and σ_y) is shown in Fig. 1a, where $z = 0$ is located at the mid-plane of the original plate. The upper surface of the original plate is defined as $z = z_o$ and the upper surface, after a layer is removed, is defined as $z = z_1$. The residual principal stress, σ_x at $z = z_1$, can be expressed as the following:

$$\sigma_x(z_1) = \frac{E}{6(1 - \nu^2)} \left\{ (z_o + z_1)^2 \left[\frac{d\phi_x}{dz}(z_1) + \nu \frac{d\phi_y}{dz}(z_1) \right] + 4(z_o + z_1)[\phi_x(z_1) + \nu\phi_y(z_1)] - 2\int_{z_1}^{z_o} [\phi_x(z) + \nu\phi_y(z)]\, dz \right\} \quad (1)$$

where ϕ_x and ϕ_y are the measured curvatures after the top[2] layer is removed. The expression for residual stress, σ_y at $z = z_1$, is obtained by interchanging the subscripts x and y in Eq 1. If there is initial curvature in the plate, the initial curvature must be subtracted from the measured curvatures ϕ_x and ϕ_y.

Figure 1b illustrates the experimentally determined curvature ϕ_x plotted against the experimental depth z_1; a plot of ϕ_y versus z_1 should be prepared as well. As illustrated in Fig. 1b, the values of $d\phi_x/dz(z_1)$ and $\int\phi_x dz$ can be obtained, directly from the graph of ϕ_x and z_1, for subsequent use in Eq 1. Values of $d\phi_y/dz(z_1)$ and $\int\phi_y dz$ are obtained in a similar manner.

[2] Top implies the surface where layers are first removed as illustrated in Fig. 1a. Identifying surfaces as top and bottom does not imply a difference between the surfaces.

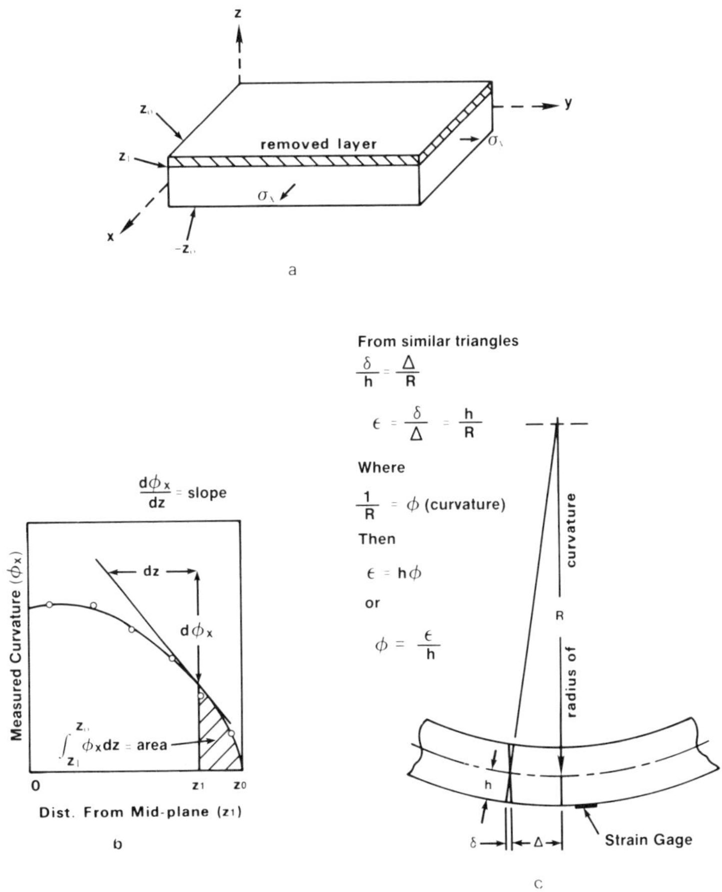

FIG. 1—(a) *Test specimen showing layer to be removed and principal stress components, according to Treuting and Read [4].* (b) *Measured curvature versus distance from mid-plane.* (c) *Curvature determined via electrical resistance strain gage measurements.*

Values of z_1 are readily determined with a micrometer; however, direct measurements of curvature are not as simple and it is frequently convenient to obtain curvatures ϕ_x and ϕ_y using electrical resistance strain gages. Strain gage measurements, obtained from conventional three-gage rosettes (either rectangular or delta configuration) installed on the bottom of the plate, provide sufficient data to determine the curvature induced principal strains. The directions of the principal strains are easily established as well. Figure 1c illustrates the geometric relationship between radius of curvature, R, and the bending strain, ϵ:

$$\epsilon = h/R \tag{2a}$$

Curvature is defined as the reciprocal of the radius of curvature; therefore:

$$\phi = \epsilon/h \tag{2b}$$

and the curvature in any direction can be established from the strain-gage-rosette measurements and the plate thickness at any stage of the layer removal process.

Hole-Drilling Strain-Gage Method

Kirsch's early theoretical work [*6*] can be used to express the radial strain, ϵ_r, relieved at a point, *P*, when a small hole is drilled through a thin sample containing a residual biaxial plane-stress field as shown in Fig. 2:

$$\epsilon_r = \sigma_x(A + B\cos 2a) + \sigma_y(A - B\cos 2a) \tag{3}$$

where

$$A = -\frac{1+\nu}{2E}\cdot\frac{1}{r^2},$$

$$B = -\frac{1+\nu}{2E}\left(\frac{4}{1+\nu}\cdot\frac{1}{r^2} - \frac{3}{r^4}\right), \text{ and}$$

$$r = R/R_o \text{ (per Fig. 2)}$$

Equation 3 can be written three times to express the strains at the mid-point of each grid of a three-element rosette. Solving the three equations simultaneously, one obtains the principal stresses σ_1 and σ_2 and their orientation, β, with respect to the rosette axes, as functions of the measured rosette strains ϵ_1, ϵ_2, and ϵ_3. Thus:

$$\sigma_1 = \frac{\epsilon_1 + \epsilon_3}{4A} - \frac{\sqrt{2}\sqrt{(\epsilon_1 - \epsilon_2)^2 + (\epsilon_2 - \epsilon_3)^2}}{4B}$$

$$\sigma_2 = \frac{\epsilon_1 + \epsilon_3}{4A} + \frac{\sqrt{2}\sqrt{(\epsilon_1 - \epsilon_2)^2 + (\epsilon_2 - \epsilon_3)^2}}{4B} \tag{4}$$

$$\beta = \frac{1}{2}\tan^{-1}\frac{\epsilon_3 - 2\epsilon_2 + \epsilon_1}{\epsilon_3 - \epsilon_1}$$

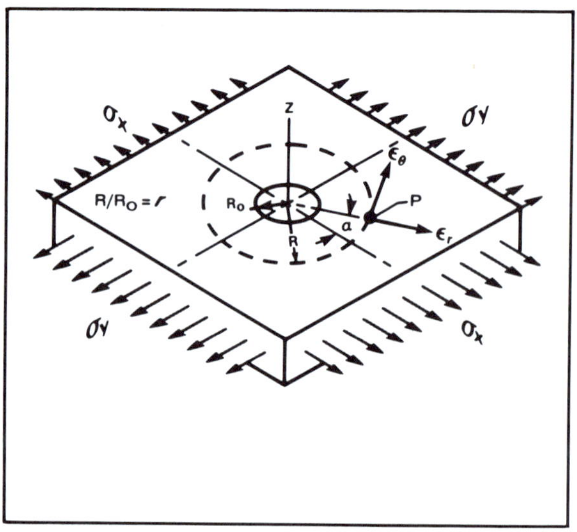

FIG. 2—*Relieved strain at (P) for a through hole in a biaxial stress field.*

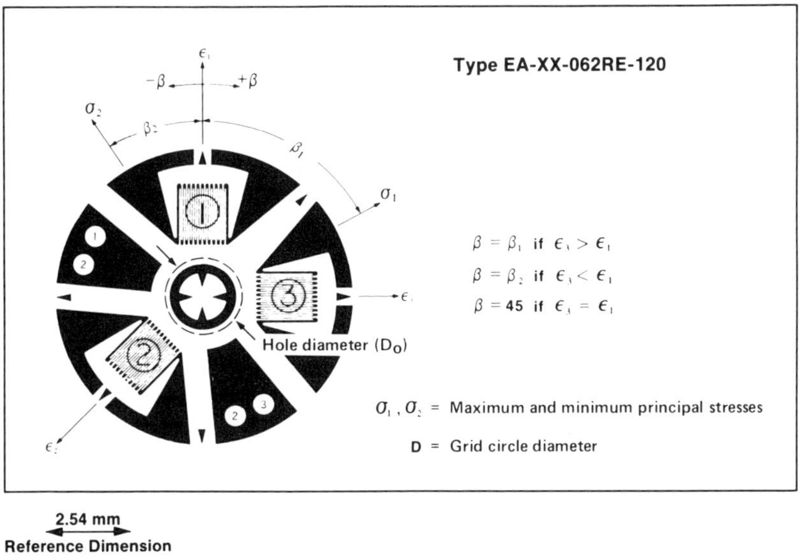

FIG. 3—*Residual stress rosette per Rendler and Vigness [7].*

The idealized coefficients A and B in the foregoing equations are strictly valid only for a thin specimen that satisfies the conditions assumed in the derivation: a through-drilled hole, uniform stress throughout the specimen thickness, specimen boundaries remote from the hole, and strain gages of zero gage length. Clearly, these conditions are not rigorously satisfied in practical applications of the hole-drilling method. In most instances, the hole-drilling method is used on thicker sections and the hole is blind rather than through. Fortunately, it has been demonstrated [7] that Eq 3 also describes the stress field around a blind hole; and further, coefficients A and B can be readily determined by experimental calibration [7].

Schajer [8], in his finite-element analysis, redefined the Kirsch coefficients as the following:

$$A = -\frac{1 + \nu}{2E} \cdot \bar{a}$$
$$B = -\frac{1}{2E} \cdot \bar{b} \tag{5}$$

where coefficients $\bar{a}$ and $\bar{b}$ are effectively independent of the specimen's mechanical properties (E and ν). Further, Schajer's coefficients ($\bar{a}$ and $\bar{b}$) account for the strain gage averaging effects (its finite dimensions). It must be acknowledged that Schajer's coefficients are applicable only to the rosette geometry established by Rendler and Vigness [7],[3] and as illustrated in Fig. 3. Schajer's coefficients apply to those residual stress fields that are uniform with hole depth and the ASTM Standard Method for Determining Residual Stresses by the Hole-Drilling Strain-Gage Method (E 837-85) includes them for use when the drilled hole depth exceeds the hole diameter by approximately 20 %.

[3] These geometric requirements are satisfied by the RE and RK residual stress rosettes manufactured by Measurements Group, Inc., Raleigh, North Carolina.

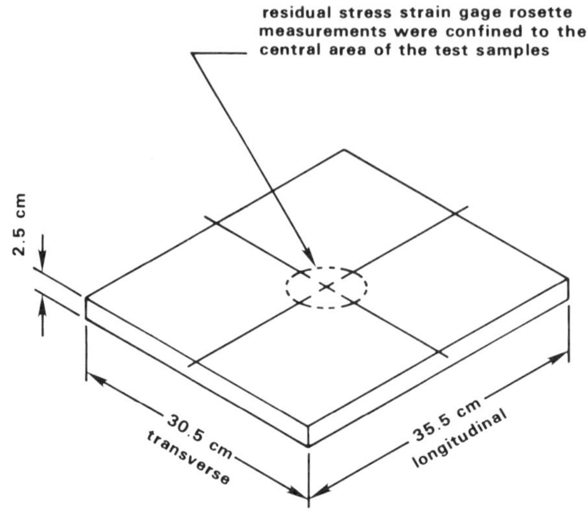

FIG. 4—*Aluminum alloy test sample, nominal dimensions.*

Test Specimens

The aluminum alloy test specimens were prepared by an industrial manufacturer and supplier of aluminum alloy products. Further, the chemical composition of the alloy is proprietary and consequently it carries no standard code or commercial identification. All heat treatments, cold rolling and stretching procedures associated with production of the plate test samples were performed by the manufacturer, who also conducted all layer-removal residual-stress analyses; these test results were made available to the author only after the hole-drilling strain-gage measurements and calculations were complete.

Layer removal and strain-gage hole-drilling procedures were used to measure residual stresses after three different thermomechanical treatments, hereinafter referred to as Condition 1, Condition 2, and Condition 3, as follows:

Condition 1—Solution heat treatment, cold water quench, room temperature age.

Condition 2—Condition 1 followed by 11.5 % reduction by cold rolling.

Condition 3—Condition 2 followed 1.25 % cold stretch.

Figure 4 illustrates the plate test specimens that were used during the strain-gage hole-drilling measurements. One test sample was provided for each of the three conditions listed above. Three Micro-Measurements EA-13-062RE-120 residual stress strain gage rosettes (geometry as in Fig. 3) were installed in the central area of each specimen as illustrated in Fig. 4. The central holes were introduced and strain measurements were made in accordance with the ASTM E 837-85. Alignment and drilling were accomplished using a commercially available milling guide equipped with an ultra-high-speed air turbine and carbide cutter.[4] Figure 5 shows the milling guide with the air-turbine assembly installed.

The concept of stretching aluminum alloys in order to reduce residual stress was recognized

[4] RS-200 Milling Guide manufactured by Measurements Group, Inc., Raleigh, North Carolina.

and employed by Kelsey [9] in his early strain-gage hole-drilling investigation of nonuniform residual stresses. Figure 6 is included to illustrate the stretch phenomenon. Figure 6a shows an assumed uniaxial residual stress field (before loading) in a rectangular bar where the tensile residual stress at the center (69 MPa) is one-half the assumed compressive residual stress at the two outside edges (−138 MPa). As axial load is applied to the bar, it is reasonably assumed that the resulting axial strain is uniform throughout the bar. Figure 6b shows typical stress versus strain responses for the outside edges (residual compression), center (residual tension), and the two points of assumed zero residual stress. At low values of applied strain, all three curves are linear and elastic with the locations of residual tensile stress yielding first, followed by the locations of zero residual stress, and, finally, yielding occurring at the locations of residual compressive stress. As shown in Fig. 6b, loading must continue to an applied strain magnitude sufficient to produce yielding of the compressive stresses. The stress-versus-strain relations are linear during unloading and it is clear that, after complete unloading, magnitudes of the ending residual stresses are only a very small fraction of the values before loading.

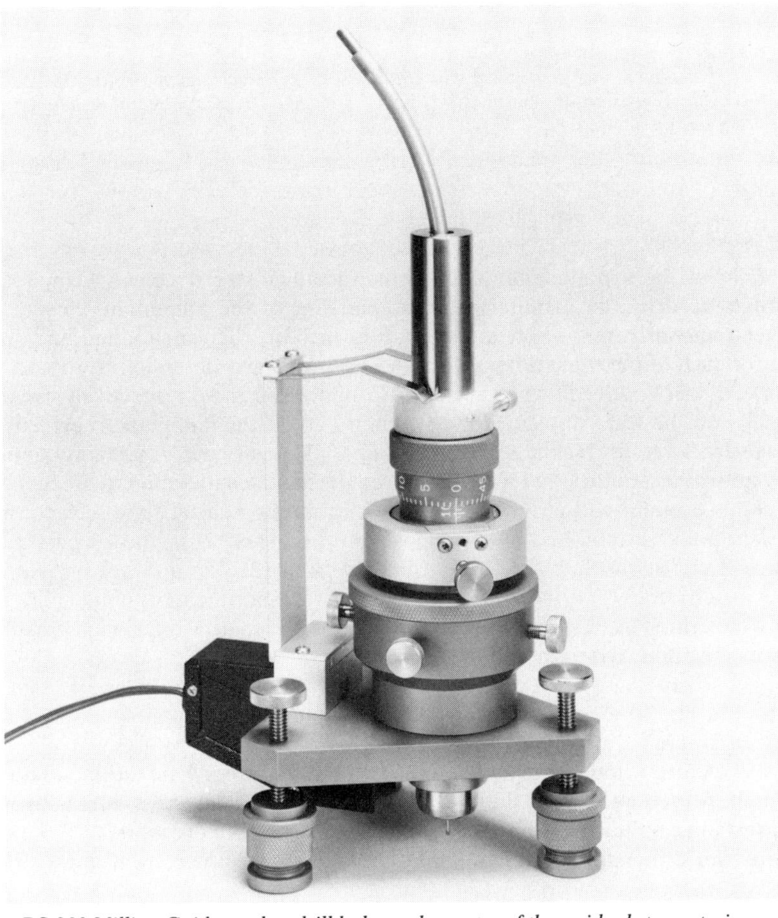

FIG. 5—*RS-200 Milling Guide used to drill holes at the center of the residual stress strain gage rosettes.*

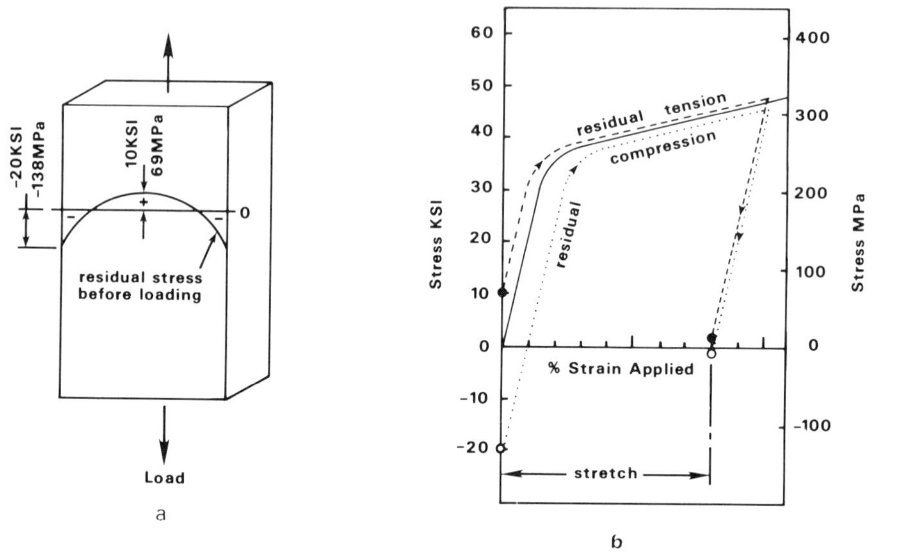

FIG. 6—(a) *Uniaxial stretch on typical aluminum alloy, assumed uniaxial residual stress condition: compression at edges, tension at center.* (b) *Uniaxial stretch on typical aluminum alloy, stress/strain response during tensile loading.*

Test Results

Layer Removal Method

Figure 7 shows the experimentally determined residual-stress-versus-specimen-thickness profiles established by the manufacturer and supplier of the aluminum alloy test plates. These layer removal results show stress profiles in both the longitudinal and transverse directions for each of the three different conditions previously described. It should be noted that the directions of both rolling and stretch, Conditions 2 and 3 respectively, were aligned longitudinally on the test samples. Good symmetry about the mid-plane is evident in all of the residual-stress-versus-thickness graphs of Fig. 7. This, of course, sustains confidence in the layer-removal techniques that were employed during the experimental process. Further, the data points established by removing layers from the top surface are seen to mesh well with the data points established by removing bottom layers. This, too, is particularly encouraging in view of the fact that layer removal from the top and bottom surfaces are performed on different test plates. Extrapolating the experimental points (Fig. 7) to the outside surfaces provides a quantitative appreciation of the surface residual stresses. These stress ranges are summarized in Table 1.

Hole-Drilling Strain-Gage Method

An RS-200 Milling Guide (Fig. 5), equipped with a high-speed air turbine and tungsten-carbide cutter, was used to drill the holes in the three residual-stress strain-gage rosettes that were installed on each of the three different test plates. The nominal hole diameter was 1.8 mm (0.072 in.) and the longitudinal distance between holes was approximately 25 mm (1 in.).

All holes were drilled using the ASTM Method E 837-85 as a guideline. Graphs of percent-relieved-strain-versus-normalized-hole-depth were plotted in the manner prescribed in Sec-

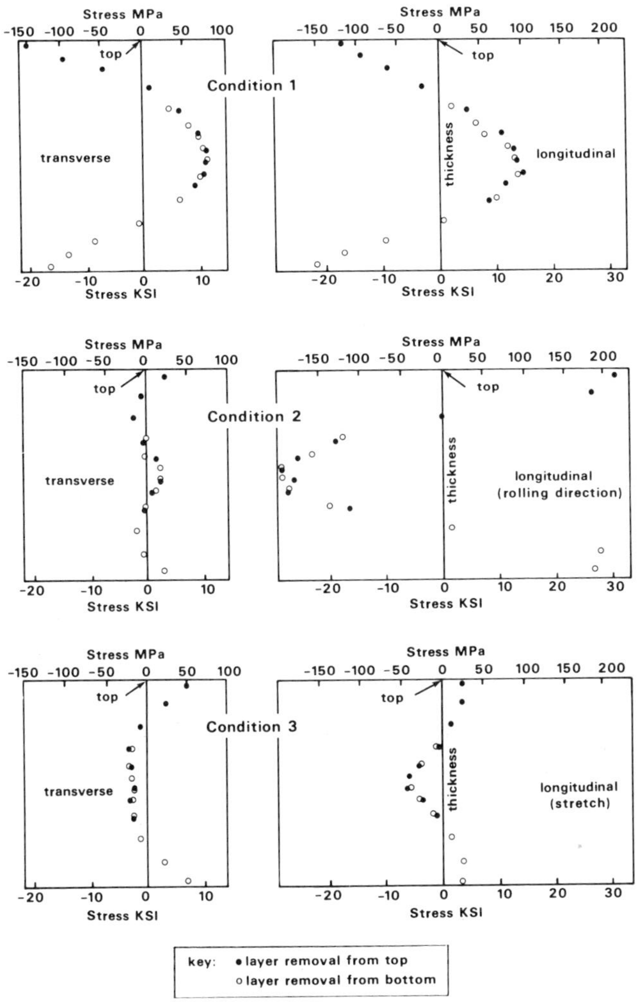

FIG. 7—*Residual stress profiles through thickness determined using layer-removal method.*

tion 8 of ASTM E 837-85. All data reduction (in accordance with ASTM E 837-85 and Ref [*6*]) was implemented using the System 4000 RESTRESS Program.[5] Figure 8 shows a RESTRESS computer-generated percent-strain-relieved-versus-normalized-hole-depth (z/D) graph that is in accordance with Section 8 of ASTM E 837-85. The envelopes bounded by the solid and broken line curves of Fig. 8 illustrate the ASTM scatterbands established in Section 8 and Fig. 4 of ASTM E 837-85. The data points on Fig. 8 are typical of all drilled holes and were judged to be not "substantially out of the scatterband" (see Section 8 of ASTM E 837-85). For purposes of this paper, the residual stresses were therefore assumed uniform with hole depth. Table 2 summarizes the hole-drilling residual stresses for all test conditions.

[5] System 4000 is the data-acquisition system manufactured by the Instruments Division of Measurements Group, Inc., Raleigh, North Carolina.

TABLE 1—*Surface stresses determined by extrapolating layer removal results to outer surfaces.*

Specimen Condition	Direction	
	Transverse	Longitudinal
1	−138 ± 14 MPa (−20 ± 2 ksi)	−150 ± 25 MPa (−22 ± 4 ksi)
2	+40 ± 10 MPa (6 ± 2 ksi)	+252 ± 23 MPa (36 ± 4 ksi)
3	+62 ± 8 MPa (9 ± 1 ksi)	+28 ± 8 MPa (4 ± 1 ksi)

Conclusion

The layer removal tests of Fig. 7 afford an excellent appreciation of the subsurface residual stress gradients as well as the stress magnitudes existing in all three test conditions. It has already been acknowledged that the near surface stresses, at the top and bottom, were necessarily obtained from two different test plates; further, it is only reasonable to expect some variation in the near surface stresses present in different test samples (even though they were from the same test condition). Table 1 reflects the uncertainties associated with the surface stress magnitudes determined by extrapolating the subsurface layer-removal data. Admittedly, the uncertainties are subjective and are dependent upon extrapolation procedures; however, the surface stress magnitudes are representative of the extrapolated data.

It is prudent to recognize that all residual stresses obtained via layer removal were derived from the nominal measured plate curvatures. Consequently, the calculated principal stresses are assumed to be invariant at each elevation (z_1) above the mid-plane. Stress magnitude variations, from point to point in elevation z_1, no matter how large or small, cannot be appraised. Conversely, the hole-drilling strain gage data of Table 2 reflect the principal residual stresses at the points where the holes were drilled. Each tabulated stress value is the average from three different hole locations, and the ± uncertainty is representative of the stress variation among the three holes on that particular test specimen. However, these

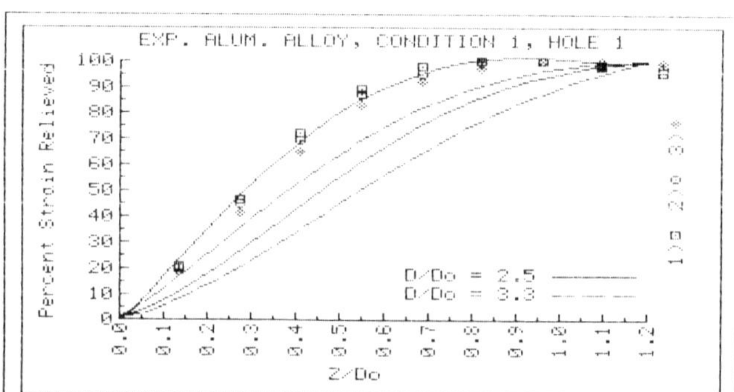

D = Diameter of gage circle
Do = Diameter of drilled hole

FIG. 8—*Computer-generated plot of relieved strains versus normalized hole depth in accordance with ASTM E 837-85, generally typical of all holes.*

TABLE 2—*Surface stresses determined by the hole-drilling method* (*ASTM E 837-85*).

Specimen Condition	Direction	
	Transverse	Longitudinal
1	−207 ± 20 MPa (−30 ± 3 ksi)	−165 ± 20 MPa (−24 ± 3 ksi)
2	+103 ± 7 MPa (+15 ± 1 ksi)	+235 ± 7 MPa (+34 ± 1 ksi)
3	+35 ± 7 MPa (+5 ± 1 ksi)	+48 ± 14 MPa (+7 ± 2 ksi)

stresses were all calculated using the concepts of ASTM E 837-85 and, consequently, they reflect stresses that are assumed to be the following:

- Uniform through the full hole depth (nominally 2 mm); and
- Less than one-half of the yield stress.

The *longitudinal* surface residual stresses established by both layer removal and hole-drilling (Tables 1 and 2, respectively) compare very favorably for all three specimen conditions. However, the *transverse* residual stress comparison, particularly for specimen Condition 1, is less favorable. Here the hole-drilling results of Table 2 (−207 ± 20 MPa) are approximately 50 % greater than the transverse stresses obtained by layer removal of Table 1 (−138 ± 14 MPa). As shown in Fig. 8, the measured strains for Condition 1 lie on the extreme edge of the ASTM E 837-85 scatterband and the simplifying assumption of uniform stress for the full hole depth may, in this case, be an over simplification.

In summary, the independent test results of the layer-removal and strain-gage hole-drilling residual stress analyses are seen to be corroborative. The advantage of the layer removal method lies chiefly with its ability to generate a complete understanding of through thickness residual stress magnitude and stress gradient. This advantage is hampered by the extensive and tedious effort required to implement the procedure and also the complete destruction of the test sample. The hole-drilling strain-gage method requires only a small fraction of the time, effort, and cost associated with layer removal; however, test results are mostly related to surface stresses and cannot be extrapolated to the inner layers of thick plates. The hole-drilling method is quite versatile, though, and is commonly used in practical field applications where layer removal procedures are entirely impractical.

The effectiveness of stretching, as a method of reducing residual stresses, is corroborated by both the layer-removal and the hole-drilling methods.

References

[*1*] Treuting, R. G., Lynch, J. J., Wishart, H. B. and Richards, D. G., *Residual Stress Measurements,* The American Society for Metals, Cleveland, OH, 1952.

[*2*] Osgood, W. R. Ed., *Residual Stresses in Metals and Metal Construction,* Reinhold Publishing Corporation, NY, 1954.

[*3*] Almen, J. O. and Black, P. H., *Residual Stresses and Fatigue in Metals,* McGraw-Hill Book Company, NY, 1963.

[*4*] Treuting, R. G. and Reed, W. F. Jr., "A Mechanical Determination of Biaxial Residual Stresses in Sheet Materials," *Journal of Applied Physics,* Vol. 22, No. 2, Feb. 1951, pp. 130–134.

[5] Measurements Group, Inc., "Measurement of Residual Stresses by the Hole-Drilling Strain-Gage Method," *Measurements Group Tech Note TN-503-2,* Measurements Group, Inc., Raleigh, NC, 1986.

[6] Timoshenko, S. and Goodier, J. N., *Theory of Elasticity,* McGraw-Hill Book Company, NY, 1963, p. 80.

[7] Rendler, N. J. and Vigness, I., "Hole-drilling Strain-Gage Method of Measuring Residual Stresses," *Proceedings,* SESA, Vol. XXIII, No. 2, 1966, pp. 577–586.

[8] Schajer, G. S., "Application of Finite Element Calculations to Residual Stress Measurements," *Journal of Engineering Materials and Technology,* Vol. 103, 1981, pp. 157–163.

[9] Kelsey, R. A., "Measuring Non-uniform Residual Stresses by the Hole-drilling Method," *Proceedings,* SESA, Vol. XIV, No. 1, 1956, pp. 181–194.

Y. Altschuler, T. Kaatz, and B. Cina[1]

Relief of Residual Stresses in a High-Strength Aluminum Alloy by Cold Working

REFERENCE: Altschuler, Y., Kaatz, T., and Cina, B., **"Relief of Residual Stresses in a High-Strength Aluminum Alloy by Cold Working,"** *Mechanical Relaxation of Residual Stresses, ASTM STP 993*, L. Mordfin, Ed., American Society for Testing and Materials, Philadelphia, 1988, pp. 19–29.

ABSTRACT: A systematic study was performed on the effectiveness of plastic deformation in tension and compression relief of residual stresses derived from the rapid quenching of 7075 aluminum. Maximum stress relief was observed after about 1.4 % deformation in tension and 1 % deformation in compression. Complete stress relief could not be obtained from either mode of deformation. Stress relief was found to be more effective after deformation in tension than in compression. Significant differences in types and levels of residual stresses were found in specimens deformed seemingly uniform in compression when measured on a free surface, or on one in contact with the platen applying the compressive force. The effect of artificial aging as a supplement to mechanical stress relief also was investigated. The practical implications of the results, particularly with respect to forgings, are discussed. Recommendations are made for further work.

KEY WORDS: residual stresses, stress relieving, plastic deformation, aluminum alloys

Introduction

Residual stresses of a very significant level are introduced into thick products made of aluminum alloys when the latter are water quenched from the temperature of solution treatment. The stresses result from the differences in cooling rate throughout the thickness of the material.

For the high-strength alloys of the 2000 or 7000 type, the method of stress relief by thermal treatment, commonly employed on steels, cannot be used since the temperatures required for effective stress relief are higher than those used for the artificial aging of these materials [*1*]. Stress relief at the actual artificial aging temperatures gives reductions of only up to about 10 to 20 % [*1*].

Stress relief is achieved in a material when the yield stress is exceeded by the residual stress. If this cannot be done by thermal means, where essentially the yield stress is reduced to a level less than that of the residual stress, then the method generally employed is to deform the material plastically. The latter treatment is almost invariably performed by the manufacturers of the raw stock, plate, bars, extrusions, forgings, etc. Since this is a vital part of the manufacturing procedure, this is perhaps one of the reasons for the dearth of publications on the subject.

For plates, bars, extrusions, and likewise simple symmetrical shapes, the plastic deformation for stress relief is performed in tension for reasons of practical convenience and

[1] Senior metallurgist, senior physicist, and chief metallurgist, respectively, Materials and Process Engineering, Engineering Division, Israel Aircraft Industries Ltd., Lod, Israel.

attainable uniformity. Van Horn [1] states that the maximum rate of stress relief is achieved in the first 0.5 % of permanent set and is virtually complete at about 2 %. In the latter case, a maximum range of residual stress of about 42 MPa (6 ksi) was found in 2014-T6 aluminum.

For hand forgings of symmetrical pancake shape, compression is the practical method of plastic deformation. According to Van Horn [1] the major extent of stress relief is achieved, as in tension, after 0.5 % permanent set, but 4 % is required for substantially complete stress relief in the transverse direction. The trouble arises when plastic deformation has to be applied uniformly to a die forging of complex shape. Most of the major plants dealing with the forging of aluminum alloys have two alternative methods of coping with this problem.

In one method, the die forging in the solution treated and quenched condition is returned to the die in which it was hot forged and is plastically deformed cold. Since the aluminum alloy forging and the alloy steel die are now both cold, there are strong possibilities of local mismatch between them. The net result is that not only will the plastic deformation obtained be nonuniform but also that some regions of the forging may experience tensile or compressive deformation. The forging after subsequent artificial aging will be defined as being in the TX54 temper, where X represents the entire heat treatment schedule performed.

The alternative procedure is to achieve the plastic deformation in dies especially designed for the purpose. In this process the plastic deformation is predominantly compressive, although not necessarily uniform, and is certainly less uniform than tensile deformation. A forging so stress relieved and subsequently artificially aged is defined as being in the TX52 temper, with X the same as previously defined. Some die forgings have shapes that allow them to be stress relieved by stretching. Variations in thickness throughout the forging may well result in nonuniform plastic deformation.

Since so relatively little systematic work seems to have been done or at least published on the effect of plastic deformation on the relief of residual stresses in high-strength aluminum alloys in general, and on forgings made from these alloys in particular, such a study was undertaken by us.

It is to be noted that in many cases in the industry, the effectiveness of a given cold working procedure is assessed by the extent to which the article so treated distorts on machining. The experience so depended upon would thus seem to be in part empirical. Systematic work on the effect of plastic deformation on stress relief should eventually facilitate standardization that would be to the benefit of all concerned. Van Horn [2] has given detailed examples of distortion due to the presence of residual stresses during machining.

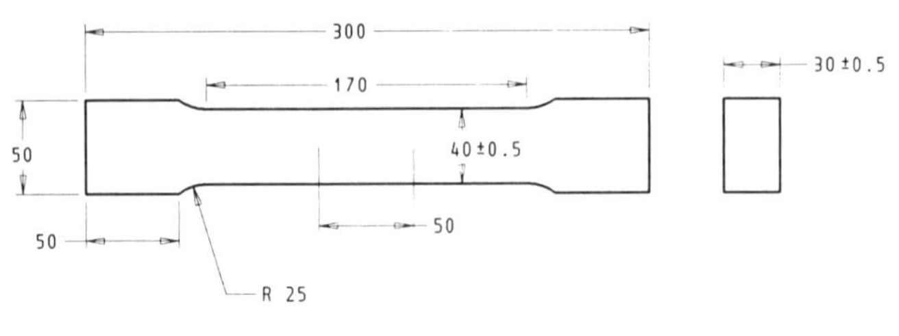

Dimensions in mm.

FIG. 1—*The configuration of specimens cold worked in tension.*

Barker and Turnbull [3] have given numerous data on the levels of residual stresses developed on the surfaces of cylindrical aluminum alloy forgings both when these were quenched in the unbored state and also when they were bored prior to solution treatment and quenching. The shapes of these forgings however clearly prevented the use of mechanical means of stress relief.

Experimental Procedure

All the work was carried out on specimens machined from a 31.8-mm (1.25-in.) thick plate of 7075 aluminum supplied in the T651 temper and conforming with the U.S. Federal Specification on Aluminum Alloy 7075, Plate and Sheet (QQ-A-250/12). Its nominal chemical composition in weight percent was as follows:

Si	0.4, max
Fe	0.5, max
Cu	1.2/2.0
Mn	0.3, max
Mg	2.1/2.9
Cr	0.18/0.28
Zn	5.1/6.1
Ti	0.2, max
Others	0.05/0.15

The temper designation T651 implies that the material had been solution treated, stress relieved by stretching, generally in the range 1.5 to 3 % permanent set at room temperature, and finally artificially aged at an elevated temperature. Since it was intended to measure the effectiveness of cold working procedures on stress relief, all specimens machined from this plate had to be re-solution treated to restore the level of residual stress associated with the particular thickness of the plate and the quenching procedure adopted previously by the manufacturer.

Specimens for cold-working experiments in compression were machined to dimensions of approximately 100 by 100 by 30 mm, the smaller dimension being taken in the thickness of the plate. The ends of the specimen that were to be in contact with the platens were parallel to within 0.0006 mm/mm. A flat-shaped specimen was deliberately chosen to simulate as closely as possible major sections in a die or hand forging. Dimensions greater than 100 mm in the plane of the specimen would have been used if a hydraulic press of capacity greater than that of the one used had been available. This point is to be emphasized since it implies that the dimensions of the specimen were not determined by considerations of determination of strength as, for example, those recommended in the ASTM Test Method for Compression Testing of Metallic Materials at Room Temperature (E 9–81). However, since only small amounts of plastic deformation were to be applied, less than 5 %, the extent of non-uniformity of deformation was not expected to be great. Specimens for cold-working experiments in tension were prepared as shown in Fig. 1. It will be seen that the thickness of the gage length in this specimen is approximately 30 mm, and therefore virtually identical to the thickness of the compression specimen. This was done to ensure, as far as possible, that both types of specimens would have the same level of residual stress prior to the cold-working experiments.

After machining, all the specimens for cold-working experiments were solution treated by heating to 465 ± 5°C, holding for 2 h at this temperature, and quenching in cold water.

After quenching, specimens were either deformed within 2 h or were refrigerated at −30°C, generally for a period up to about 24 h to prevent any aging or stress relief. The cold-working experiments were carried out within this 24-h period. Measurements of the residual stresses developed on quenching from the temperature of the solution treatment were also made within this 24-h period and of course prior to any cold working experiments.

Residual stresses were measured primarily by back reflection X-ray diffraction techniques, using the accepted two-exposure film method and employing Cu Kα X-radiation. Measurements were made on the diffraction ring from the (511)/(333) planes of the aluminum alloy specimens and using pure nickel powder on the surface of the specimen for purposes of calibration. The specific surfaces examined in both types of specimen will be described later. The accuracy of the measurements was ± 21 MPa.

For the cold-working experiments in compression a hydraulic press of 150 ton capacity was employed. The aluminum alloy specimens were compressed at a rate of about 2 %/min between two hardened low-alloy steel platens machined to a surface roughness of 63 rms. Individual specimens were compressed in the direction of their smallest dimension, approximately 30 mm, to permanent sets of approximately 0.5, 1, 2, 3, and 5 %.

A corrosion preventive oil served as lubricant between the platens and the specimen. The alignment of the specimen and platens was such that after maximum plastic deformation of 5 %, the maximum variation in the thickness of the specimen was less than 0.2 %. The extent of barrelling was very slight in all the samples and was greatest in that compressed by 5 %. Its maximum extent was 0.2 %.

For the cold-working experiments in tension, the specimens were extended at a rate of about 4 %/min in an Instron Model tensile testing machine of 50 ton capacity. Prior to extension, parallel lines were scribed in the central portion of the gage length of each specimen, perpendicular to its length and 50 mm apart. The precise distance between the two scribe marks was measured prior to extension and after full unloading to an accuracy of ±0.01 mm. As for compression, sometimes successive trials were required to obtain the amount of permanent extension required. Separate specimens were plastically deformed 0.3, 1.4, 2.7, 2.9, and 4.2 %, respectively. Calibration of the tensile test machine showed that a small bending effect operated on the specimen such that a bending stress of a magnitude

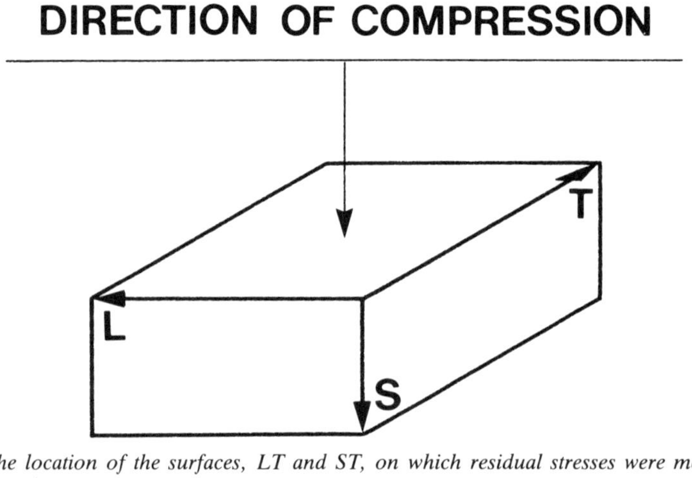

FIG. 2—*The location of the surfaces, LT and ST, on which residual stresses were measured in the compression specimens.*

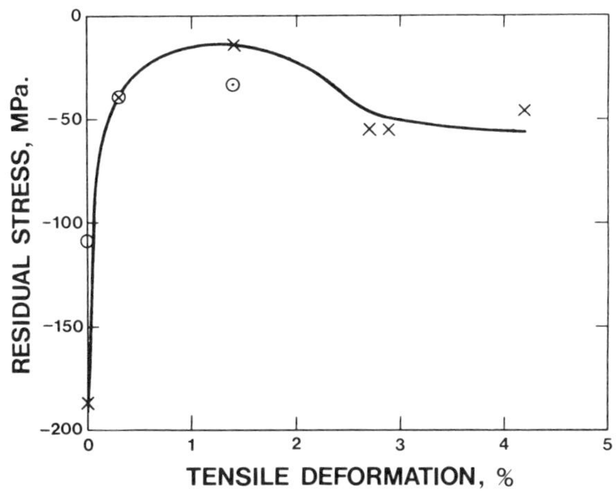

FIG. 3—*Residual stresses in specimens deformed plastically in tension. X, as deformed; O, as deformed and aged to T73 temper.*

up to 3 % of that of the applied tensile stress could be experienced by the surface of the specimen subsequently to be examined for residual stresses.

The residual stresses in specimens deformed in compression were measured on two locations as indicated in Fig. 2. One is the LT surface in immediate contact with a steel platen in the hydraulic press and the other, ST, is one of the free surfaces adjacent and perpendicular to the first. In each case the residual stress was measured at points close to the center of each face.

For specimens deformed in tension, the residual stress was measured on the broadest side of the gage length, that of 40-mm initial dimension. For the tensile specimen and the LT face of the compression specimen, the residual stress was measured in the original longitudinal direction of the plate. For the ST face of the compression specimen, the residual stress was measured in the original transverse direction of the plate. Thus in the specimen cold worked in tension, the tensile force acted parallel to the plane in which the residual stress was measured, while in the specimen cold worked in compression, the compressive force acted both parallel and perpendicular to the planes in which residual stresses were measured since two surfaces were examined, the ST and LT ones, respectively, as in Fig. 2.

Since it was of interest to know to what extent the residual stresses in the specimens would be reduced simply by the final overaging stage of heat treatment, this treatment was applied to a specimen in the as-quenched condition and to several others plastically deformed in tension. The specific details of this heat treatment were as follows:

1. 7 hours at 108°C.
2. 27 hours at 163°C.

Results

Specimens Deformed in Tension

The results of measurement of residual stress in specimens deformed plastically in tension are shown in Fig. 3. It will be seen that there is a high residual stress of 186 MPa in compression present in a specimen as quenched from the temperature of solution treatment.

Cold working by 1.4 % reduces the level of residual stress to only about −14 MPa. Further cold work in tension up to 4.2 % seems to increase the level of residual compressive stress, however this latter effect is uncertain due to the experimental error of the residual stress measurements. The effect of artificial aging after cold working is discussed later.

Specimens Deformed in Compression

The results of measurement of residual stresses in specimens deformed plastically in compression are shown in Fig. 4. The results are presented for measurements made on the LT and ST surfaces, with LT and ST being as depicted in Fig. 2. The LT surface was in contact with the cold-working steel platens.

Note first that the levels of residual stress for these LT and ST surfaces in specimens as-quenched from the temperature of solution treatment are high, −207 and −165 MPa respectively, and are of the same order of magnitude as that found in the tensile specimen in the same condition of heat treatment, −186 MPa. This similarity is in keeping with the similar thickness of the two types of specimen. Since the residual stresses in the compression specimen operate in the plane of the surface examined, the possibily somewhat higher value of residual stress in the LT surface as compared with that in the ST surface, if significant, probably results from the faster quenching rate of the bulk of the material behind the ST surface as compared with that behind the LT surface. This of course is the classical end effect and results in a smaller difference in cooling rate between the ST surface the the bulk material behind it. This in turn will lead to a lower residual stress on the ST surface. The value of residual stress of −207 MPa measured on the major surface of the compression specimen of 30 mm thickness is in keeping with values of about −262 to −277 MPa measured

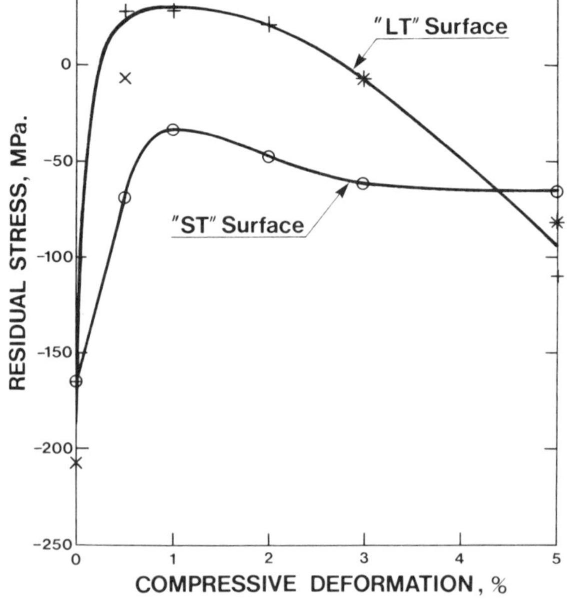

FIG. 4—*Residual stresses in specimens deformed plastically in compression. O, ST surface; X, LT surface as deformed; +, LT surface as deformed and chemically milled 0.1 mm.*

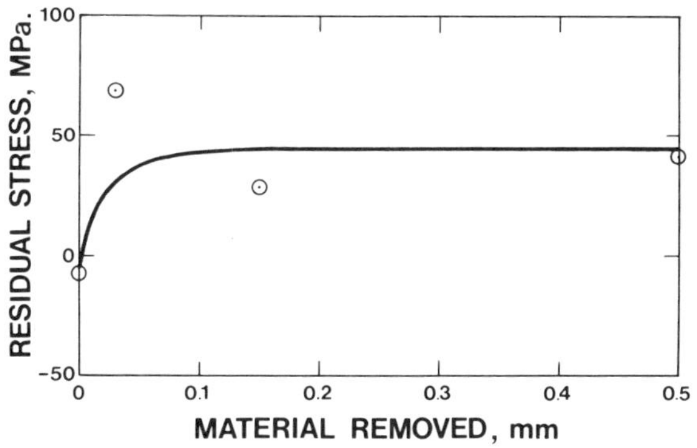

FIG. 5—*The effect of chemical milling on the level of residual stress on the LT surface of a specimen plastically deformed 0.5 % in compression.*

in the center of the major surface of plates of 7075 aluminum of the dimensions 400 by 400 by 70 mm quenched in cold water at 20°C after solution treatment at 467°C [*4*].

Boyer and Boivin [*5*] quote a value of −230 MPa for a quenched 70-mm thick 7075 aluminum plate. These higher levels of residual stress are as expected from thicker specimens.

Plastic deformation in compression caused changes in residual stress in a manner partially similar, though not identical, with that observed in tension. For the ST surface that was physically free during the compression operation, the level of residual stress progressively decreased with increase in plastic deformation up to about 1 %. Further cold work seemed to result in a slight increase in the level of compressive residual stresses although the increase observed is of the order of accuracy of the measurements. It is to be noted that the minimum level of residual stress obtained, as measured on the ST surface, was about −35 MPa.

The results of measurement of residual stress on the LT surface were significantly different from those measured on the ST surface. Since it was known that there can be considerable friction effects between the specimen and the steel compression platens such that there was some uncertainty as to the actual extent of plastic deformation obtaining on the immediate surface of the specimen, all of the specimens were chemically milled to a depth of about 0.1 mm on the LT surface prior to examination by X-ray diffraction. The chemical milling was carried out in a saturated aqueous solution of sodium hydroxide. The results of measurement of residual stress on such chemically milled specimens are shown in curve LT in Fig. 4. Some of the compression specimens were also examined for residual stress prior to chemical milling. These individual results are also shown in Fig. 4. Chemically milled specimens show very rapid relief of residual stress up to about 0.25 % plastic deformation.

Further cold work up to about 2.8 % introduces a relatively low level of residual tensile stresses. Beyond 2.8 % cold work there is a suggestion of the reintroduction of residual compressive stresses as observed in measurements on the ST surface.

That 0.1 mm depth of material was of the correct order to be removed by chemical milling was shown by an experiment whereby the LT surface of a specimen plastically deformed in compression by 0.5 % was chemically milled in stages to a depth of 0.5 mm. The level of residual stress as determined by X-ray diffraction was determined at each stage. The results are shown in Fig. 5. Although there is some scatter in the results, Fig. 5 suggests that there is little change in the level of residual stress after removal of more than 0.1 mm of material.

As will be seen in Fig. 4, chemical milling seemed to affect the results of residual stress measurement only for 0 and 0.5 % compression. Allowing for experimental error, the differences observed for these two specimens may not be significant. There is certainly no obvious explanation for the results from the specimen in the as-quenched condition. Further work is required to clarify the effect for specimens compressed between 0.5 and about 2.5 %.

Thermal Stress Relief

Of the several temperatures at which artifical aging can be performed on material made from 7075 aluminum, that associated with the T73 temperature is the highest. It was therefore of interest to know to what extent residual stresses, as initially formed or remaining after cold working, were reduced by such a thermal treatment. The results for specimens plastically deformed in tension are shown in Fig. 3.

It will be seen that for a specimen in the condition as-quenched from the temperature of solution treatment there is a significant reduction in the level of residual stress, from −186 MPa to −108 MPa, a reduction of about 40 %. The fact that Van Horn [*1*] reported reductions of up to only 20 % for 7075 aluminum would be explained by the fact that the artificial aging was achieved at 120°C (the T6 temper) whereas in the present work the aging was achieved at a maximum temperature of 163°C (the T73 temper). For specimens as-quenched and stress relieved mechanically by 0.3 and 1.4 %, respectively, no significant further stress relief was observed on aging to the T73 temper.

Summary

The high level of residual compressive stress, in the range of about 165 to 207 Mpa, found on the surfaces of specimens 30 mm thick in the condition as-quenched from the temperature of solution treatment was significantly reduced by 1.4 % plastic deformation applied in tension and 1 % in compression, respectively. Whereas the maximum amount of stress relief by cold working in tension was about 90 %, the corresponding amount for compression was about 80 % for measurements made on a surface not previously in contact with the compression platens. The implication from these results is that plastic deformation in compression is less efficient than plastic deformation in tension is in keeping with the calculated predictions of Boyer and Boivin [*5*] for a quenched 70-mm thick 7075 aluminum plate. These calculations predict that almost complete relief of residual stress should be obtained after about 2 % stretch, but that even after 4 to 6 % compression significant compressive residual stresses should be present on the major surface of the plate. No explanation is offered by Boyer and Boivin for these differences.

Further problematic points in the present work are in the observations that complete stress relief could not be obtained by cold deformation in tension or compression, and that after the maximum amount of stress relief had been obtained by deformation in compression or in tension, further deformation seemed to result in a slight increase in the residual compressive stress. This trend, however, is uncertain since it is of the order of the experimental error. It was observed for compressive and tensile deformation to the extent of only 5 and 4 %, respectively. The effect of further plastic deformation was not investigated.

Redevelopment of residual stresses on cold working would seem to imply non-uniformity in the cold-working process. This was not expected in the tensile specimen. The relative dimensions of the specimen employed are close to those in Fig. 6 of ASTM Methods of Tension Testing of Metallic Materials (E8-856) for plate type material. The length of the grips is smaller than required, and this may have introduced a small bending effect, especially at the larger extensions. This would be in addition to the small bending effect equivalent to about 3 % of the tensile load determined on calibration of the tensile machine.

In the compression specimen the nonuniformity would have been caused primarily by the friction between the platens and the specimen. Hsu [*6*] has made a theoretical and experimental analysis of the compression test for ductile materials and has pointed out how friction effects at the ends of the specimen can lead to complicated stress conditions and inhomogeneous strain, all of which will vary with the type of material and the proportions of the specimen. He shows the very beneficial effects of adequate lubrication and analyzes the stress conditions maintaining under varying degrees of lubrication.

Although the need for lubrication between the platens and the specimen is well recognized [*5,7*], frictional effects can not be completely eliminated [*6*]. Thus recent research [*8*] shows experimentally and by calculation that for amounts of plastic deformation in compression and of extent sufficient to cause cracking, different states of stress, and very significantly different levels of stress, exist at different locations in the specimen. In attempting to relate such observations to the present results, it is to be noted that a relatively dead zone with respect to the extent of deformation is observed adjacent to the platens while the external surface of the specimen near the equator shows a combination of axial and hoop strain [*8*]. Likewise, the zone of maximum deformation was found to be in the center of the specimen [*8*]. It is known that shear stresses operate in the zone immediately adjacent to the platens and this has recently been dealt with quantitatively by Dadras and Thomas [*9*]. Papirno et al. [*10*] studied the axial compression of cylinders of different alloys including aluminum particularly with respect to the onset of cracking. Aluminum alloys 6061-T651 and 7075-T6 began to show surface wrinkling, the orange peel effect, on the previously smooth cylindrical surface of the specimens, at 5 to 7 % height reduction. On subsequent deformation, when barrelling developed, shear cracks formed in these wrinkled equatorial regions. In the present work, the extent of deformation was insufficient to lead to wrinkling or cracking.

Although several investigators [*8,9,10*] were interested primarily in states of stress and strain at large extents of compressive deformation, their results can perhaps be interpreted to indicate the possible onset of nonuniformity of the extent of plastic deformation; and the modes of stress obtaining, in different regions of the specimens for nominal amounts of compression between 1 and 5 % as in the present work. It is of interest that in Hsu's work [*6*], an unlubricated copper specimen showed signs of barrelling very early in the deformation, at about 6 % compression by our interpretation of his data. While the present work is also concerned with such relatively small extents of deformation, the specimens were lubricated so that in fact only slight barrelling was observed. The latter results are in keeping with Papirno's statement [*7*] that the effect of friction on stress and strain distributions is of consequence only when the deformations are on the order of 10 % or more.

Since in the present work a different response to cold working was observed on measurement of residual stress on the two surfaces of the compression specimens examined, the most logical explanation is that such measurements are a sensitive indication of the inhomogeneity of stress and strain in those regions, the inhomogeneity resulting from friction.

In practice, forgings have complex shapes, so that contact between their surfaces and those of the dies in which they are cold worked will be less ideal than those obtaining in symmetrical laboratory specimens. Furthermore, the surface finish of the forgings, and probably also of the dies, will also be rougher than those in laboratory specimens and platens. Accordingly, a greater extent of inhomogeneity is to be expected in practice because of both the geometry of the set up and because of friction.

It is of interest to note the results of mechanical stress relief obtained by the Aluminum Company of America [*11*] on 101.6-mm (4-in.) thick hand forgings of 7079 aluminum. After full heat treatment to the T6 temper but without mechanical stress relief, the maximum residual stress on the surface of the forgings was about −17 ksi (−117 MPa). It possibly would have been somewhat higher in the as-quenched condition. Mechanical stress relief in tension and compression, albeit by unstated amounts, resulted in maximum residual stresses on the surface of the forgings of about 2.5 ksi (17 MPa) and 6 ksi (41 MPa), respectively. That is, the residual stress was changed from compression to tension in both cases, and more efficient stress relief was observed in tension than in compression as in the present work. Since the residual stresses after compression were also measured through the thickness of the forgings, the residual tensile stresses observed were clearly measured on the surface directly subjected to the compressive force and are therefore in keeping with the present results, curve LT in Fig. 4. All the present results for compressive stress relief show that it may be important to examine both a directly compressed and a free surface.

Conclusions

The practical implications of the present results are several.

1. Cold working to about 0.5 % in tension or compression relieves the major extent of residual stresses in 7075 aluminum in the as-quenched condition.

2. Maximum stress relief was obtained after about 1.4 % deformation in tension and 1 % in compression.

3. Mechanical stress relief in tension is to be preferred to that in compression where this is practicable, for reasons of effectiveness; about 90 % as against about 80 %, respectively.

4. Even for articles such as hand forgings of pancake form deformed seemingly uniformly in compression, considerable variation in the type and level of residual stress may be present throughout the bulk. For articles of complex form such as die forgings whose mechanical stress relief is carried out in the original hot forging die, even greater variation in the type and level of residual stress is to be expected throughout the bulk. The work reported herein should be extended to articles of different thickness and from different aluminum alloys to establish the generality of the conclusions reached. It is also desirable to measure the gradients of residual stresses especially in articles stress relieved by plastic deformation in compression.

References

[1] Van Horn, K. R., *Aluminum, Vol. III, Fabrication and Finishing*, American Society for Metals, Metals Park, OH, 1967, Chapter 10, pp. 355–382.

[2] Van Horn, K. R., "Residual Stresses Introduced During Metal Fabrication," *Journal of Metals, Transactions*, American Institute of Mining and Metallurgical Engineers, Vol. 197, 1953, pp. 405–422.

[3] Barker, R. S. and Turnbull, G. K., "Control of Residual Stresses in Hollow Aluminum Forgings," *Metal Progress*, American Society for Metals, Nov. 1966, pp. 60–65.

[4] Jeanmart, P. and Bouvaist, J., "Finite Element Calculation and Measurement of Thermal Stresses in Quenched Plates of High Strength 7075 Aluminum Alloys," *Materials Science and Technology*, Vol. 1, Oct. 1985, pp. 765–769.

[5] Boyer, J. C. and Boivin, M., "Numerical Calculations of Residual Stress Relaxation in Quenched Plates," *Materials Science and Technology*, Vol. 1, Oct. 1985, pp. 786–792.

[6] Hsu, T. C., "A Study of the Compression Test for Ductile Materials," *Materials Research and Standards*, Vol. 9, No. 12, Dec. 1969, p. 20.

[7] Papirno, R. in *Metals Handbook, Vol. 8, Mechanical Testing*, American Society for Metals, 1985, pp. 55–58.

[8] Mescall, J., Papirno, R., and McLaughlin, J., "Stress and Deformation States Associated with Upset Tests in Metals," *Compression Testing of Homogeneous Materials and Composites, ASTM STP 808*, American Society for Testing and Materials, 1983, pp. 7–23.

[9] Dadras, P. and Thomas, J. F., "Deformation Inhomogeneities in Upset Forging," *Compression Testing of Homogeneous Materials, ASTM STP 808*, American Society for Testing and Materials, 1983, pp. 24–39.

[10] Papirno, R., Mescall, J. F., and Hansen, A. M., "Fracture in Axial Compression Tests of Cylinders," *Compression Testing of Homogeneous Materials, ASTM STP 808*, American Society for Testing and Materials, 1983, pp. 40–63.

[11] Aerospace Technical Information Bulletin, "Alcoa Stress-relieved Forgings," Aluminum Company of America, Series 68, No. 2, 1968.

DISCUSSION

N. W. Hung[2] (*written discussion*)—Dr. Cina brought up an important issue of reducing the residual stress by coldworking. I understand that your main concern was the dimensional stability rather than the benefit of the compressive surface stress. Your data as the result of the uniaxial tension or compression on the simple geometry specimen, is very beneficial to help us understand the technical aspect of this complicated subject; nevertheless, I am a little concerned about the general practicality of this method. For very complex geometry, high precision (tolerance $<\pm 0.0005$ in.), or smaller size parts, the technique of cold working to reduce the residual stresses can become very expensive, if not impossible. I think the alternative technique that is worth considering is the thermal-mechanical, uphill quenching technique which can help reduce the stress due to solution heat treat quenching.

Y. Altschuler, T. Kaatz, and B. Cina (*authors' closure*)—The results of the work were intended to be applied primarily to forgings and specifically thick forgings subsequently to be heavily machined to parts for structural purposes. Such parts, even those of complex geometry, are routinely cold worked to reduce their residual stresses. We would agree that if special cold-working dies have to be designed and manufactured because of the complexity of the part, this entails a considerable expense, however even this is done if uniform and maximum dimensional stability on subsequent machining is essential. Uphill quenching is indeed one alternative method for reducing residual stresses resulting from quenching from an elevated temperature. The process has not yet been generally accepted nor standardized although it is now being more thoroughly evaluated.

[2] Hewlett Packard, Santa Rosa Division, 1400 Fountain Grove Parkway, Santa Rosa, CA 95404.

R. H. Leggatt and T. G. Davey[1]

Measurements of the Reduction Due to Proof Loads of Residual Stresses in Simulated Pressure Vessel Welds

REFERENCE: Leggatt, R. H., and Davey, T. G., **"Measurements of the Reduction Due to Proof Loads of Residual Stresses in Simulated Pressure Vessel Welds,"** *Mechanical Relaxation of Residual Stresses, ASTM STP 993,* L. Mordfin, Ed., American Society for Testing and Materials, Philadelphia, 1988, pp. 30–41.

ABSTRACT: An investigation of the effects of proof loading on residual stresses in test panels simulating welds in ammonia storage spheres was performed. The two test panels contained welds with different degrees of misalignment and distortion. Measurements were made of the surface residual stresses before and after loading to about two thirds yield and of the strain changes during loading. Previous research has suggested that the residual stress after proof loading is a function of material yield strength and applied proof stress only. In the present project, it was found that the relaxation of residual stresses transverse to the weld was sensitive to the local geometry. However, a simple method for including the effects of residual stresses in defect assessments is proposed.

KEY WORDS: residual stresses, proof loads, mechanical relaxation, pressure vessels, misalignment, defect assessment, stress concentration factors.

Nomenclature

E	Young's modulus
$_L$	Longitudinal
SCF	Stress concentration factor
$_T$	Transverse
$\Delta\epsilon$	Strain change
$\Delta\sigma$	Stress change
ϵ_1	Strain at maximum load
ϵ_2	Strain after unloading
ν	Poisson's ratio
σ_a	Applied stress
σ_P	Pressure test stress
σ_r	Residual stress before loading
σ_r'	Residual stress after loading
σ_S	Service stress
σ_{YS}	Yield strength

[1] The Welding Institute, Abington Hall, Abington, Cambridge, CB1 6AL UK.

Introduction

NOTE—Please note that throughout this paper, the directions "transverse" and "longitudinal" or "parallel" are relative to the direction of welding unless otherwise stated.

The application of tensile stresses to welded structures causes a reduction in the peak residual stresses in the welds. This phenomenon has been known for many years and is discussed in review papers by Burdekin [*1*] and Nichols [*2*]. Burdekin points out that if a structure is loaded with the same pattern of stresses as is to be applied in service, then the structure itself will find the residual stress areas that need attention, and the treatment will ensure that such stresses are reduced in critical areas. This would usually be the case in a hydrotest of a pressure vessel. Nichols quotes from the work of Nordell and Hall [*3*], who worked on 25-mm thick A212A mild steel plate, and found that the mean value of the residual longitudinal stress was reduced from the as-welded value by an amount equal to the prestress. This result was in accordance with a simple explanation of the phenomena based on linear-elastic-perfectly-plastic material behavior. It suggests that the residual stress σ_r' in weld metal of yield stress σ_{YS} subjected to applied stress σ_a would be given by the following:

$$\sigma_r' = \sigma_{YS} - \sigma_a \tag{1}$$

However, closer examination of the results of Nordell and Hall [*3*] shows that, although the average longitudinal residual stress in the weld after prestressing was reduced in accordance with Eq 1, the peak residual stress after loading was significantly higher than the predicted value.

Data from a variety of published sources [*3–6*] are summarized in Fig. 1, normalized with respect to weld metal yield strength. The data from Nordell and Hall [*3*] and Kihara et al. [*4*] relate to welded mild steel or carbon manganese steel plates. The data from Jesensky and Vargova [*5*] were deduced from measured residual stresses and calculated applied stresses in 120-mm thick pressure vessels; the weld metal yield stress was not quoted but was estimated to be equal to the maximum measured as-welded longitudinal residual stresses.

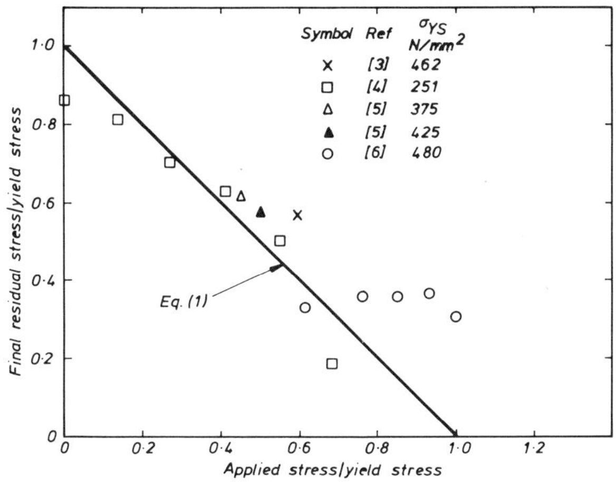

FIG. 1—*Variation of measured residual stresses after proof loading with applied stress.*

The data from Potter and Millard [*6*] were obtained on shot-peened aluminum alloy bar: all stresses were compressive.

It can be seen on Fig. 1 that much of the data lie significantly above the line corresponding to Eq 1. In fact, it was noticable that test procedures which measured highly localized residual stresses tended to give higher values, while those that measured bulk residual stresses over larger volumes gave results closer to Eq 1.

The present investigation was part of a project whose aim was to determine allowable defect sizes in spherical ammonia storage vessels that were subject to proof testing before entering service. In order to maximize the allowable defect sizes, it was required to make an allowance for the mechanical relaxation of the residual stresses in the welds due to the application of the proof loads. In view of the uncertainty over the effect of prestressing, it was decided to make some measurements of residual stresses and stress changes during loading in test panels containing welds simulating those in the storage vessels. It was suspected that the deviations from Eq 1 might be associated with stress concentrations, that would modify the local value of the applied stress. Hence, the test panels were made with misalignments and distortions representing those that might occur in practice.

Experimental Procedures

Specimens

The test panel material was 13-mm thick BS 1501-224-490B LT 50 carbon manganese steel of yield strength 415 N/mm^2 and tensile strength 545 N/mm^2. The welds were deposited with

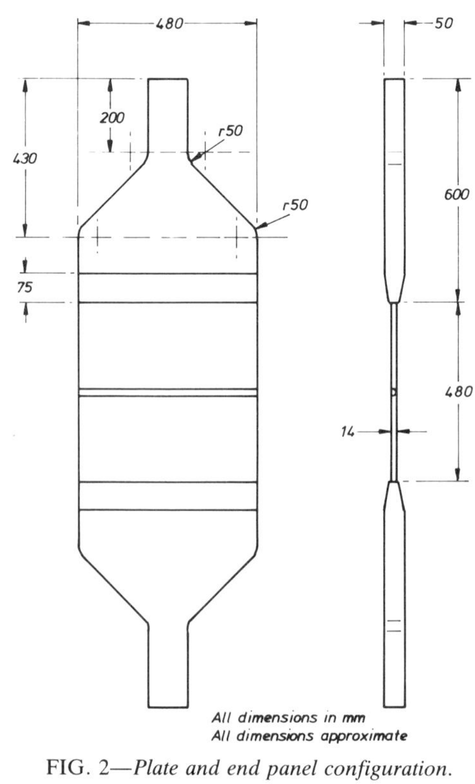

FIG. 2—*Plate and end panel configuration.*

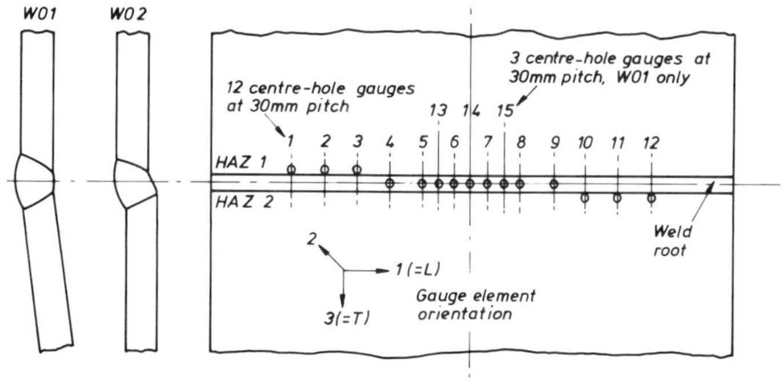

FIG. 3—*Center-hole rosette gage locations, with enlarged end view showing angular distortion and misalignment.*

Fortrex 35A electrodes. The tensile properties of the welds were measured as yield strength 405 N/mm^2 and tensile strength 527 N/mm^2 on all-weld metal samples.

There were two test panels, each containing a central weld (Figs. 2 and 3). Specimen W01, referred to as the "aligned" panel, was 480 mm wide (along the weld), 500 mm long and 13.65 mm thick. It had a small hi-low misalignment of 0.1 mm at the weld, and an angular distortion about the weld of 1.45°. Panel W02, the "misaligned" panel, was 480 mm wide, 435 mm long and 13.71 mm thick. It had a misalignment of 1.7 mm and an angular distortion of 0.38°. The root passes of the welds in the specimens were hand-ground to a smooth profile, as would be the case in a pressure vessel. The test panels were welded to thicker end panels to facilitate tensile loading as shown in Fig. 2.

Residual Stress Measurements

The residual stresses were measured at selected locations in the heat-affected zones and welds using the center-hole rosette gage method. The technique used and the derivation of the results were as described by Beaney [7], who quotes an overall accuracy of ±8 % for stresses up to 65 % of yield. The holes were formed using a rotating jet of abrasive powder, and were approximately 2 mm in depth and diameter. The method gives a weighted average of the residual stresses over the depth of the hole, and is mainly influenced by the stresses in the region between 0.2 and 1.2 mm from the surface.

Sets of four center-hole measurements were used to characterize the residual stresses in the test plates in each condition of interest; before loading, after loading transverse to the weld, and (for panel W01 only) after loading in the longitudinal direction, parallel to the weld. The gage locations are shown in Fig. 3. Each set of four gages was consisted of two located on the centerline of the weld root and two located in the heat-affected zone (HAZ) regions on either side of the weld root. Three additional measurements were made on panel W01 after the longitudinal loading.

Loading Procedure

The plates were instrumented with conventional uniaxial and biaxial strain gages in order to monitor the strain changes during loading. The gage locations are shown in Fig. 4. Gages were applied at the same locations on both surfaces of both plates. The biaxial gages (elements 1 and 2) were used to monitor the far-field strains. The uniaxial gages were used to monitor the strains in the vicinity of the welds. The layout of the uniaxial gages was

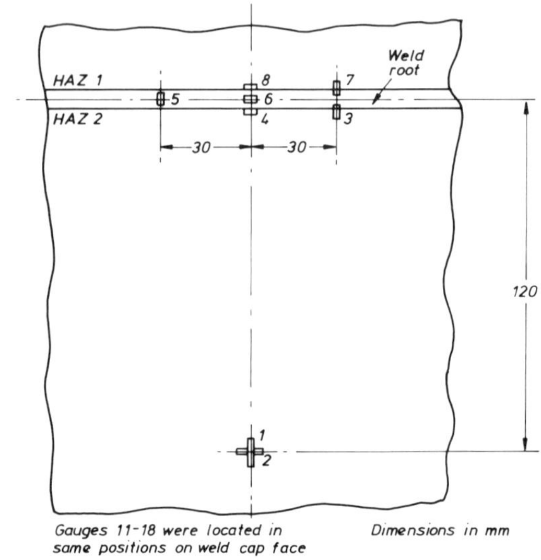

FIG. 4—*Layout of uniaxial and biaxial gages to monitor strain changes during loading.*

chosen to allow them to be used later for internal residual stress measurement by the block removal method, though this was not in fact carried out. All gages were located at least 15 mm from the site of the nearest center-hole gage, to avoid stress concentration effects due to the 2-mm diameter holes.

The plates were loaded in a 1.8 MN test machine. The aligned plate W01 was loaded first in the transverse direction, perpendicular to the weld and subsequently (after residual stress measurement, removal of the thicker end panels, and reattachment in the appropriate configuration) in the longitudinal direction, parallel to the weld.

Plate W02 was loaded in the transverse direction only. Each plate was loaded to a mean applied stress of 230 N/mm^2, using a force of approximately 1.5 MN.

Results and Discussion

Surface Residual Stresses

The surface residual stress measurements are summarized in Table 1. It can be seen that the longitudinal stresses in the plates as-received (before loading) were tensile, in the range from +62 to +326 N/mm^2, and the transverse stresses were low, in the range from −34 to +107 N/mm^2.

The changes in residual stresses measured at similar locations before and after loading transverse to the welding direction are given in Table 1. The net changes after loading in both directions are also given for plate W01. There was very little consistency among the stress changes at the four types of location and in fact every set of four stress changes had one value whose sign was opposite to those of the others. It should be noted, however, that residual stresses cannot be measured at exactly the same location more than once using the center-hole technique. The locations of the gages used to measure the stresses after loading were 30 or 60 mm displaced from those used before loading. Hence, the inconsistency in the stress changes may represent scatter in the spatial distribution of residual stresses rather than in the effect of the loading.

A more consistent picture emerges from the average residual stresses and stress changes listed in the right hand column of each part of Table 1. The transverse stresses in both plates showed a small decrease after transverse loading (−53 and −23 N/mm²). The longitudinal stresses also showed a small decrease after transverse loading (−31 and −33 N/mm²). The most significant average stress change was that occurring in the longitudinal direction in plate W01 after transverse and longitudinal loading (−153 N/mm²). The transverse stresses on the same plate appeared to show very little net change after dual loading (−7 N/mm²), though this average value was heavily influenced by one large positive reading (126 N/mm²) in the weld metal after longitudinal loading. Additional readings were made on weld metal (after loading) at adjacent locations 13 to 15, and these showed a mean transverse residual stress of −52 N/mm², giving a net change of −66 N/mm² compared with the original mean stress level. The original results, excluding the value of +126 N/mm², gave a mean transverse stress of −33 N/mm² and a net change of −47 N/mm². The reading of +126 N/mm² after dual loading appears to be unrepresentative.

A prediction of the maximum stress in a particular direction after loading in that direction can be made using Eq 1. Taking the weld and parent yield strengths of 405 and 415 N/mm² respectively, and the measured mean applied stress of 269 N/mm² (see next section), the predicted maximum stresses after loading are 136 N/mm² in weld metal and 146 N/mm² in HAZ. All the measured residual stresses were less than, or not more than 5 % greater than, these values. However, the initial transverse residual stresses in the panels were low (maximum 107 N/mm²), and hence the tests did not demonstrate whether the higher transverse residual stresses that might be present in a pressure vessel would be relieved in accordance with Eq 1.

TABLE 1—*Measured residual stresses, N/mm².*

Gage locations	HAZ 1	Weld	Weld	HAZ 2	Average
Before loading	#1	#4	#7	#10	
After T-load	#2	#5	#8	#11	
After L-load	#3	#6	#9	#12	
1a: W01, aligned plate, longitudinal stresses (parallel to weld):					
Before loading	99	326	267	318	252
After T-load	158	216	240	272	222
After L-load[a]	153	97	130	19	100
Change after T-load	+59	−110	−27	−46	−31
Total change	+54	−229	−137	−153	−153
1b: W01, aligned plate, transverse stresses (perpendicular to weld):					
Before loading	−29	−4	−20	107	14
After T-load	−27	−85	−45	1	−39
After L-load[b]	2	−54	126[c]	−47	7
Change after T-load	+2	−81	−25	−106	−53
Total change	+31	−50	+146	−154	−7
1c: W02, misaligned plate, longitudinal stresses (parallel to weld):					
Before loading	62	307	271	210	213
After T-load	102	254	190	173	180
Change after T-load	+40	−53	−81	−37	−33
1d: W02, misaligned plate, transverse stresses (perpendicular to weld):					
Before loading	−34	11	16	27	5
After T-load	−41	−34	−55	59	−18
Change after T-load	−7	−45	−71	+32	−23

[a] Longitudinal stresses at locations 13 − 15 after L-load were 84, 66, 84 (av. 78).
[b] Transverse stresses at locations 13 − 15 after L-load were −55, −76, −24 (av. −52).
[c] Suspect reading, see text.

Stress Changes During Load Cycles

During the load cycle, individual locations in the vicinity of the weld were expected to deform plastically under the combined effects of residual and applied loading, but then to unload elastically when the load was removed. Hence, it is not possible to deduce the overall stress changes from the measured strain changes due to the possibility of plasticity. However, because the unloading phase may be assumed to have been elastic, it is possible to calculate the decrease in stress during unloading. If it is further assumed that the stress at maximum load was less than or equal to the yield strength, it is then possible to deduce the maximum stress after loading.

The stress changes during unloading were deduced from the measured strain changes using Hooke's law:

$$\Delta\sigma_{\mathrm{L}} = \frac{E}{(1 - \nu^2)} (\Delta\epsilon_{\mathrm{L}} + \nu\Delta\epsilon_{\mathrm{T}})$$

$$\Delta\sigma_{\mathrm{T}} = \frac{E}{(1 - \nu^2)} (\Delta\epsilon_{\mathrm{T}} + \nu\Delta\epsilon_{\mathrm{L}})$$

Subscripts L and T refer to the directions along and transverse to the welding direction, respectively. Longitudinal and transverse strain readings were taken from pairs of gages at equal distances from the weld centerline, such as gages 3 and 4 on Fig. 4.

The measured strains at maximum load, ϵ_1, and after unloading, ϵ_2, and the deduced stress changes are given in Table 2.

The shape of the plate in the region of each gage location is given in the final column of Table 2, and may be related to the sketches of the weld cross-section given in Fig. 3. The following descriptions of shape are used:

Flat—Parent plate remote from weld. All regions are described as flat with respect to loading parallel to the weld, that is, there is no stress concentration in this direction.

Concave—Root and both adjacent HAZs on W01 (due to angular distortion). HAZ1 on root side and HAZ2 on cap side on W02 (due to misalignment).

Convex—Both HAZs on cap side of W01. HAZ2 on root side and HAZ1 on cap side on W02.

Inflection—Weld root on W02.

Proud—Weld cap on both plates.

Wherever the specimen was flat, the stress change in the direction of loading was expected to be roughly equal and opposite to the applied stress (230 N/mm^2). In fact they were consistently slightly higher than this value, as follows:

- W01, transverse loading, $\Delta\sigma_{\mathrm{T}} = -254, -284$; average -269 N/mm^2.
- W01, longitudinal loading, $\Delta\sigma_{\mathrm{L}} = -270, -279, -270, -262, -286, -262$; average -272 N/mm^2.
- W02, transverse loading, $\Delta\sigma_{\mathrm{T}} = -233, -291$; average -262 N/mm^2.

TABLE 2—*Strain and stress changes during unloading.*[a]

Region	Gages		ϵ_{IL}	ϵ_{IT}	ϵ_{2L}	ϵ_{2T}	$\Delta\sigma_L$	$\Delta\sigma_T$	Shape	$\Delta\sigma/\Delta\sigma_{av}$[b]
2a: W01, Aligned plate, transverse loading:										
Plate	2	1	−226	1198	175	−44	7	−254	flat	
HAZ 1	4	3	−244	3773	145	1592	−60	−470	concave	1.74
Weld root	6	5	−255	2649	145	483	−57	−465	concave	1.73
HAZ 2	8	7	−306	3592	94	1360	−61	−480	concave	1.78
Plate	11	12	−123	1412	198	67	−19	−284	flat	
HAZ 1	14	13	−399	449	−97	−104	31	−105	convex	0.39
Weld cap	16	15	−407	93	−101	−46	60	−11	proud	0.04
HAZ 2	18	17	−360	437	−66	−4	37	−80	convex	0.30
2b: W01, Aligned plate, longitudinal loading:										
HAZ 1	4	3	1108	−739	−193	−358	−270	−2	flat	1.00
Weld root	6	5	1541	86	212	432	−279	−12	flat	1.04
HAZ 2	8	7	1242	−951	−59	−566	−270	−1	flat	1.00
HAZ 1	14	13	1207	−432	−62	−42	−262	2	flat	0.97
Weld cap	16	15	2453	220	1092	571	−286	−13	flat	1.06
HAZ 2	18	17	1524	−617	251	−216	−262	4	flat	0.97
2c: W02, Misaligned plate, transverse loading:										
Parent	2	1	−381	1095	−4	−44	8	−233	flat	
HAZ 1	4	3	−515	4272	−126	2106	−59	−466	concave	1.78
Weld root	6	5	−491	1823	−106	220	−21	−338	inflection?	1.26
HAZ 2	8	7	−428	935	−31	−43	24	−195	convex	0.74
Parent	11	12	−357	1325	−4	−60	−14	−291	flat	
HAZ 1	14	13	−415	562	−43	−89	40	−122	convex	0.47
Weld cap	16	15	−407	616	−50	−50	36	−127	proud	0.48
HAZ 2	18	17	−360	1180	−66	−476	−41	−340	concave	1.30

[a] All strains are microstrains, all stresses N/mm^2.

[b] $$\frac{\Delta\sigma}{\Delta\sigma_{av}} = \frac{\text{stress change in loading direction}}{\text{mean stress change in flat regions } (-269 \text{ N/mm}^2)} = \text{experimental SCF.}$$

It is considered likely that these higher than expected levels of stress were due to non-uniform loading across the width of the specimen: all measurements were taken on or near the loading axis where the stress would be expected to be highest. These stresses have an overall average of -269 N/mm^2, a value with which the unloading stresses at other locations may be compared.

Wherever the specimen was concave, the unloading stress change in the direction of loading was higher than the mean value of -269 N/mm^2:

- W01, transverse loading, $\Delta\sigma_T = -470, -465, -480$ N/mm^2.
- W02, transverse loading, $\Delta\sigma_T = -466, -340$ N/mm^2.

A high value (-338 N/mm^2) was also found at the root of the weld on W02, that has been described as a point of inflection, though the precise shape at this point was not determined accurately.

Low values of unloading stress change were found where the shape was convex, or on weld beads that were proud of the surface:

- W01, transverse loading, $\Delta\sigma_T = -105, -11, -80$ N/mm^2.
- W02, transverse loading, $\Delta\sigma_T = -195, -122, -127$ N/mm^2.

The stress changes in the direction normal to the loading direction were positive in regions where the dominant stress change was less than average (ranging from $+24$ to $+60$ N/mm^2); they were small in the flat regions (ranging from -19 to $+8$ N/mm^2); and they were negative where the dominant stress change was above average (ranging from -21 to -61 N/mm^2).

The ratio (stress change in the direction of loading)/(mean flat stress change, -269 N/mm^2) is given in the final column of Table 2.

These ratios are in effect experimental stress concentration factors (SCF), and can be seen to be greater than unity in concave regions and less than unity in convex or proud regions. In the extreme case of the weld cap on W01, the SCF was almost zero, indicating that this region was not exposed to any significant stresses during loading or unloading.

Discussion of Other Factors that may affect Residual Stresses after Proof Loading

In the present tests, transverse and longitudinal loads were applied to the test panels separately. In an actual vessel, biaxial loads would be applied simultaneously. It was noted in the present project that there were some biaxial effects under uniaxial loading: for example, the mean longitudinal residual stresses decreased by about -30 N/mm^2 during transverse loading. Additionally, the sign of the longitudinal stress changes during removal of transverse loads appeared to be a function of the magnitude of the transverse stress changes at similar locations. However, these biaxial interaction effects were small compared with the dominant stress changes, and it is considered likely that biaxial interactive effects would also be small under simultaneous biaxial loading.

Another factor that might affect the final residual stresses in a vessel is overmatching or undermatching of the yield strength of the weld metal relative to that of the parent steel. In the present study, the yield strengths of the two materials were similar (405 and 415 N/mm^2) and no effects of mismatch were expected or observed. If a significant degree of mismatch occurred, then it is likely that longitudinal residual stresses would be a function of the yield strength of the region of interest, while transverse residual stresses would be controlled by the lesser of the yield strengths of the two materials. These predictions are largely speculative, however, and further investigation of the effects of mismatch and of simultaneous biaxial loading would be desirable.

Significance of Results with Respect to Defect Assessment

The position is normally straightforward in the case of defects that are oriented transverse to the weld and subject to longitudinal stresses. Since the longitudinal residual stresses are always high in the as-welded condition, yielding can occur during proof loading. Since the SCFs in the longitudinal direction are approximately equal to unity, the final residual stress is given by the following:

$$\sigma_{rL}' = \sigma_{YS} - \sigma_{PL} \tag{2}$$

where σ_{PL} is the applied stress longitudinal to the weld during proof loading. In the case of a pressure vessel, the proof loads would be applied by means of a pressure test.

The position is more complex with respect to defects lying parallel to the welding direction and subject to transverse stresses. The as-welded residual stresses were found to be low (-4 to $+107$ N/mm^2). However, these were measured on panels that were unrestrained at the time of measurement. In the actual structure, the welds would be subject to restraint, and higher residual stresses could be present. Hence, once again, it is necessary to assume that the final stresses are controlled by unloading from yield:

$$\sigma_{rT}' = \sigma_{YS} - \text{SCF} \cdot \sigma_{PT} \tag{3}$$

where

σ_{PT} = applied stress transverse to the weld during pressure testing.

The problem here is that the SCF may not be known. Values between 0.74 and 1.78 were observed in the present project at locations on the root side of the weld, and values between 0.04 and 1.30 at locations on the cap side. Unless the weldment geometry were controlled in some way, there is no guarantee that more extreme values could not be found in service. The lower the values of SCF occurring, the smaller the beneficial effect of pressure testing and the greater the final residual stress predicted by Eq 3.

It is also the case that regions with low SCFs will be subject to low service stresses, provided that the stresses due to the service loads are similar in distribution and lesser in magnitude than the pressure test stresses. The total stress σ_{tot} under service stress σ_{ST} (after pressure testing, and assuming that no reverse yielding occurs) is given by the following:

$$\sigma_{tot} = \sigma_{YS} - \text{SCF} \cdot (\sigma_{PT} - \sigma_{ST}) \tag{4}$$

Since the pressure test stress is greater than the service stress, the quantity within parentheses in Eq 4 is always positive, and the total stress must therefore be less than yield, regardless of the value of the SCF. Hence, it would appear to be reasonable to assess the safety of defects on the basis of the assumption of total stress (residual plus applied) equal to the yield strength. This represents a significantly less severe assumption than that currently applicable to as-welded structures, for which the total stress would be equal to (σ_{YS} + $\sigma_S \cdot$ SCF), as for example in the British Standards Institution document on Defect Acceptance Levels (BSI PD6493:1980).

The same assumption would also be conservative with respect to stresses parallel to the welding direction, and it may be simpler to carry out all assessments on the basis of this simple rule.

It must be emphasized that these recommendations only apply to structures in which the stress distribution during proof testing is similar in distribution to that during service loading. The authors have recently encountered a situation where this was definitely not the case. This concerned a long cylindrical gas storage vessel supported in two cradles with its axis horizontal. The proof testing using pressurized water imposed large bending loads on the structure that did not occur during normal service.

Conclusions

The measured surface residual stresses on the root side of the weld were fairly scattered. The longitudinal residual stresses (parallel to the welding direction) were high in the as-welded condition (average +253 N/mm^2) and showed a significant decrease after loading (average change −153 N/mm^2). The transverse residual stresses were low in the as-welded condition (average 10 N/mm^2) and showed only a small decrease after loading (average change −32 N/mm^2).

The stress changes during the unloading part of the applied load cycles were more consistent. The stress changes in the longitudinal direction during longitudinal loading were approximately equal to the applied load divided by the cross-sectional area. The stress changes in the transverse direction during transverse loading were dependent on the stress concentration factor (SCF) at the region in question: this varied from 0.04 to 0.48 on weld caps, through 0.30 to 0.74 in convex regions, and up to 1.30 to 1.78 in concave regions.

For the purpose of the assessment of defects in a vessel that has been subjected to a proof loading, it is suggested that, on the basis of the present results, the total stresses in service (that is, the sum of applied and residual stresses) may be assumed to be equal to the yield strength of the region in which the defect is located.

Acknowledgments

The authors are grateful to Imperial Chemical Industries Plc who sponsored the project and supplied the test panels.

References

[*1*] Burdekin, F. M., *Heat Treatment of Welded Structures,* The Welding Institute, 1969.
[*2*] Nichols, R. W., "The Use of Overstressing Techniques to Reduce the Risk of Subsequent Brittle Fracture—Part 1," *British Welding Journal,* January 1968, pp. 21–84.
[*3*] Nordell, W. J. and Hall, W. J., "Two Stage Fracturing in Welded Mild Steel Plates," *Welding Journal,* Vol 44, 1965, pp. 124s–134s.
[*4*] Kihara, H., Masubuchi, K., Kududa, T. and Iida, K. "Initiation and Propagation of Brittle Fracture in Residual Stress Field," International Institute of Welding Document X-219-59.
[*5*] Jesensky, M. and Vargova, J., "Calculation and Measurement of Stresses in Thick-Walled Welded Pressure Vessels," *Zvaacske Spravy/Welding News,* Vol. 31, No. 4, 1981, pp. 79–87.
[*6*] Potter, J. M. and Millard, R. A., "The Effect of Temperature and Load Cycling on the Relaxation of Residual Stresses," *Proceedings of the Syracuse Conference on Advances in X-Ray Analysis,* 1977, pp. 309–319.
[*7*] Beaney, E. M., "Accurate Measurement of Residual Stresses on any Steel using the Centre-hole Method," *Strain,* Vol 12, No. 3, July 1986, pp. 99–106.

DISCUSSION

N. W. Hung[1] (*written discussion*)—Can you suggest any analytical treatment for the magnitudes and direction of the final residual stresses after a multiaxial proof loading?

R. H. Leggatt and T. G. Davey (*authors' closure*)—The effects of a multiaxial proof loading could be analyzed using an incremental elastic-plastic analysis, incorporating an appropriate multiaxial yield criterion and plastic flow rule.

N. W. Hung (*written discussion*)—On one slide, you showed a stress-corrosion induced crack started inside a weld and propagated toward the interior. Is it in conflict with the general thinking that stress corrosion cracking is environmentally induced and usually starts at the outside of a specimen?

R. H. Leggatt and T. G. Davey (*authors' closure*)—The slide in question showed stress corrosion cracking in a valve body wall. The cracking initiated in a crevice between the inner surface of the valve body and an insert ring attached by a single fillet weld. The presence of the crevice caused an accumulation of chloride irons and a geometric stress concentration. This was a classic case of environmental and geometric factors conducive to the initiation of stress corrosion cracking.

D. J. DePaul[2] (*written discussion*)—The subject paper deals with laboratory-type test specimens. To what extent have similar type tests been made on metal steel pressure vessels which have not received a thermal stress relief but which have been exposed to cyclic temperature and pressure following proof loading to reduce residual stresses?

R. H. Leggatt and T. G. Davey (*authors' closure*)—Measurements of residual stresses in a thick-walled pressure vessel after proof loading were given in the paper by Jesensky and Vargova [5].

[1] Hewlett Packard, Santa Rosa Division, 1400 Fountain Grove Parkway, Santa Rosa, CA 95405.
[2] Plant Apparatus Division, Westinghouse Electric Corporation, P.O. Box 425, Monroeville, PA.

Vibratory Stress Relief

R. D. Ohol, B. V. Nagendra Kumar, and R. A. Noras[1]

Measurement of Vibration-Induced Stress Relief in the Heavy Fabrication Industry

REFERENCE: Ohol, R. D., Nagendra Kumar, B. V., and Noras, R. A., **"Measurement of Vibration-Induced Stress Relief in the Heavy Fabrication Industry,"** *Mechanical Relaxation of Residual Stresses, ASTM STP 993,* L. Mordfin, Ed., American Society for Testing and Materials, Philadelphia, 1988, pp. 45–57.

ABSTRACT: The use of mechanical vibrations to relieve residual stresses in engineering components is increasing in use. Several vibratory conditioning systems are commercially available. When the technique was considered for treatment of fabricated structures at Larsen and Toubro Limited, residual stress measurements were made to determine the effectiveness of the treatment. A bedplate structure was vibrated by means of a commercially available variable-frequency vibrator. Residual stresses were measured near a weld location in the structure before and after treatment. Stress relief of about 30 to 57 % was noted. During the vibratory treatment, surface strains were monitored; at resonance the applied surface strain amplitude was measured to be about 600 microstrains. Subsequent to machining, the bedplate showed good stability of dimensions. A stainless steel bowl assembly was fabricated with carbon steel cooling jackets. The bowl contained heavy weldments but could not be thermally stress relieved prior to machining. The bowl was treated by vibrations and the close tolerances on the machined dimensions were satisfied. Both the components described have been installed on site and have maintained dimensional stability.

KEY WORDS: welded joints, residual stress, vibrations, vibratory conditioning, stress relief, stress measurement

Introduction

Manufacturing processes such as casting, welding, and machining often cause a buildup of residual stresses in components. These residual stresses must be relieved because they can add to the service stresses and may increase the susceptibility of the components to failure by brittle fracture, accelerated corrosion, or stress-corrosion cracking. Residual stresses can cause unacceptable distortion in structural components such as machine frames and bedplates. For example, unstayed welded structures have been heavily distorted due to weld shrinkage [*1*]. Nonuniform cooling of heavy section castings has resulted in high levels of residual stress; when machined, these castings often change shape as stressed material is removed.

Clear guidelines as to a safe, permissible residual stress level are not easily available. Two types of service conditions or requirements are envisaged:

Dimensional stability over extended periods—As has been shown in studies on weathering of gray iron castings, a reduction in internal stress of only about 10 % is sufficient to ensure shape stability [*2*].

[1] Machinery Development, Larsen and Toubro Ltd., Powai Works, PRDH, P.O. Box 8901, Bombay, India, 400 072.

Cyclic loading or corrosion—Tensile residual stresses should be reduced to as low a level as possible. In welded structures a minimum of 80 % reduction is thought to be acceptable [*3*].

Engineering components normally have been stress-relieved by thermal means. A sufficient increase in the temperature of metallic materials causes their yield strength to decrease to very low levels. For example, at about 600°C (1100°F) the yield strength of most steels is as low as 10 MPa, or about 5 % of the room-temperature strength. All internal stresses exceeding this low level are relieved by plastic deformation. Prolonged holding at stress-relieving temperatures causes some further reduction in residual stress by time-dependent creep. In steels, 60 to 85 % of all stresses are eliminated by holding the structures at 510 to 570°C (975 to 1083°F) for one hour [*2*]. Slow cooling to room temperature is necessary to avoid generation of residual stresses due to nonuniform temperature distribution.

Stress Relieving by Mechanical Means

The objective of stress relief by plastic deformation may be achieved at room temperature by superimposing an externally applied stress on the residual stress field. The applied stress must be of a magnitude such that the algebraic sum of the combined stresses exceeds the room temperature yield strength of the material, causing plastic deformation. When the externally applied stresses are removed, the level of residual stresses will be reduced. This mechanism is the basis of vibratory stress relieving or vibratory conditioning processes.

Real-life engineering structures have complex internal stress fields. The application of loads large enough to create effective stress relieving stresses is difficult. These loads need to be applied at carefully selected locations on the structure to be effective. Because large displacements and strains are a result of a structure resonating under the influence of a time-varying force of small magnitude, controlled vibration becomes an elegant solution to the problem of loading arbitrarily-shaped structures.

Vibratory conditioning is an effective method in the treatment of structures made from the following:

Austenitic stainless steels—Stainless steels in which the precipitation of chromium carbides at thermal treatment temperatures reduces their corrosion resistance.

Dissimilar metals—Distortion due to differences in thermal expansion coefficients at elevated temperatures.

Age-hardened alloys—Alloys that may lose their strength at normal stress-relieving temperatures.

Research Findings

Many researchers have studied the effect of vibrations on residual stresses in laboratory specimens and in some welded structures. Some of the important considerations for successful application of vibratory conditioning are:

- It is not necessary to completely relieve residual stresses when distortion control is the prime consideration [*4,5*].
- It is necessary to visualize the resonance mode shapes of the structure for proper location of the supports and the vibrator. Higher modes of vibration should be excited so as to generate a more uniform strain amplitude field [*6*].

- The externally applied strain amplitude must exceed a threshold value for stress relief to occur [*7,9,10*]. A large reduction in residual stress is possible [*8,9*].

Need for Measurement of Residual Stress Relief in Structures

The research findings demonstrate the usefulness of vibratory treatments in controlling distortion in fabricated structures. At Larsen and Toubro Limited, several types of components, ranging from rotating machinery to pressure vessels, are manufactured. In components intended to serve as rigid members and supports, rather than as pressure-retaining vessels, the distortion due to fabrication of heavy sections and dissimilar metals was thought to be easily treated by vibratory methods.

Vibratory conditioning using proprietary machines is gaining acceptance by designers and fabricators alike. Although the treatment is primarily intended for shape-stabilization of structures, the effectiveness of any treatment is often judged by measuring the stress relief obtained and comparing it with the stress relief obtained by thermal means. The following sections cover the program undertaken at Larsen and Toubro Limited, Bombay.

Residual Stress Measurement Technique

The hole-drilling strain-gage method for the measurement of residual stresses, ASTM Standard Method for Determining Residual Stresses by the Hole-Drilling Strain-Gage Method (ASTM E 837-85), was selected for the following reasons:

- Ease of in-situ measurements.
- Availability of accurate strain-gage instrumentation.

Strain gages mounted on the component surface record the realignment of the surface when a small volume (1.59 mm diameter by 1.59 mm deep (1⁄16 in. diameter by 1⁄16 in. deep)) of the stressed component is removed by drilling. Use of a rosette gage allowed the determination of the orientation of the axes of the principal stresses and their magnitude.

Prior to use of the hole-drilling strain-gage method on an engineering structure, it was used to measure stresses in a test specimen (Fig. 1). Clamping at the split end of the ring allowed known levels of stresses to be developed. The ring specimen was fabricated from an IS:226 structural quality steel plate (IS Standard for Structural Steel (Standard Quality) IS:226-75) with the following properties:

- Carbon content = 0.23 %, maximum,
- Yield strength = 255 MPa, minimum,
- Ultimate tensile strength = 410 − 530 MPa, and
- Elongation = 23 %, minimum.

A residual stress gage (Type EA-06-062RE-120 from Photolastic Inc., USA) was mounted at the location shown in Fig. 1. Elements 1 and 3 of the gage were aligned so as to be radial and tangential, respectively, with respect to the ring center. A three-wire temperature-compensating circuit was used for connecting the gages to the instrumentation as required by the ASTM E 837-85. Insulation of the gages from the ground was checked. A model RS-200 milling guide (Photolastic Inc., USA) was used for hole-drilling. The tool center was aligned with a microscope. A blind hole of 1.59 mm diameter by 1.59 mm deep (1⁄16 in. diameter by 1⁄16 in. deep) was drilled with an end mill driven by a variable-speed electric drilling machine [*11*]. The D/D_0 ratio was about 3 and is within the range specified by

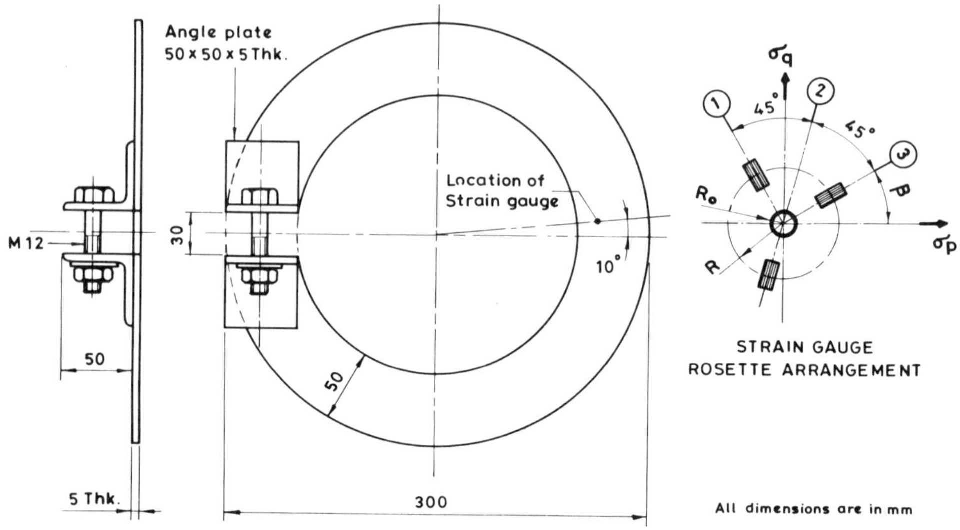

FIG. 1—*Test ring for evaluating hole-drilling technique.*

ASTM E 837-85. The principal stresses were computed from the measured strain change using the equations in ASTM E 837-85.

The stress distribution in the ring was also determined numerically by finite element analysis. The plate was modeled using 6-noded linear-strain-triangle (LST) type plane stress elements. Figure 2 shows the half-symmetry FEM model comprising of 72 elements and 185 nodes. As seen in Fig. 1, the loading bolt is above the plane of the split ring. Any possible bending effect due to such a loading arrangement was not considered in the analysis.

The numerical and strain-gage results of the analysis are presented in Fig. 3. The measured tangential stress deviates by about 10 % from the finite element result. The magnitude of both the tangential and radial stresses is observed to be greater than the numerical results. Since the strain-gage was located in the compressive region, some parasiting of the stress values during drilling probably has led to the inaccuracies.

Measurement of Residual Stress Relief on an Engineering Structure

A heavy gear box bedplate for a cement plant grinding mill was fabricated by welding of plates rather than I-beams and channels (Fig. 4). A weldable structural steel according to the IS Standard for Weldable Structural Steel (IS:2062-69) was used. The steel has the following properties:

- Composition:
  C = 0.20 % maximum,
  S,P = 0.055 % maximum,
- Yield strength = 235 MPa, minimum,
- Ultimate tensile strength = 410 – 530 MPa,
- Elongation = 23 % minimum,
- Elastic modulus = 200 GPa, and
- Poisson's ratio = 0.3.

The bedplate was expected to distort during machining of its top surface. Though not mandatory, bedplates and other fabricated structures normally have been stress-relieved by thermal means at Larsen and Toubro. Vibratory conditioning was suggested as a cheaper alternative.

Experimental Details

In order to verify the effectiveness of the technique the following experiments were planned:

1. Residual stress measurements before vibratory conditioning.
2. Measurement of dynamic strains during the treatment using a separate strain gage connected to the dynamic strain gage instrumentation.
3. Monitoring of vibration levels during the monotonic increase from rest of the vibrator frequency for identifying the resonance frequencies.
4. Residual stress measurement after the vibratory conditioning treatment.

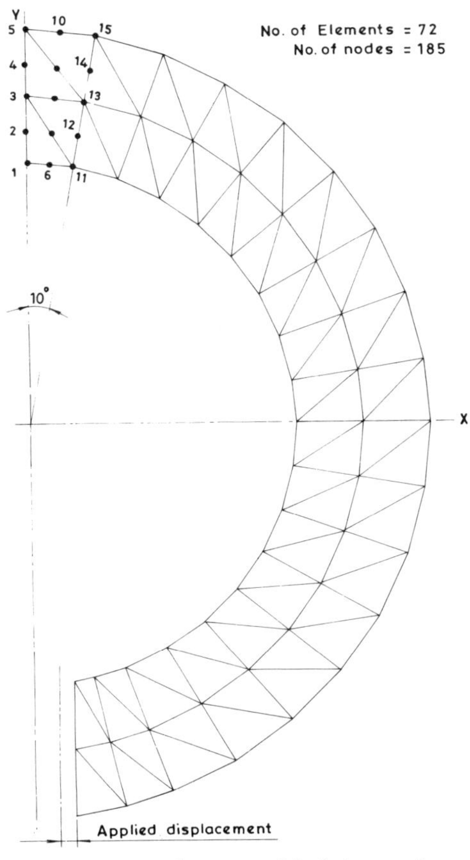

FIG. 2—*Finite element model of ring specimen.*

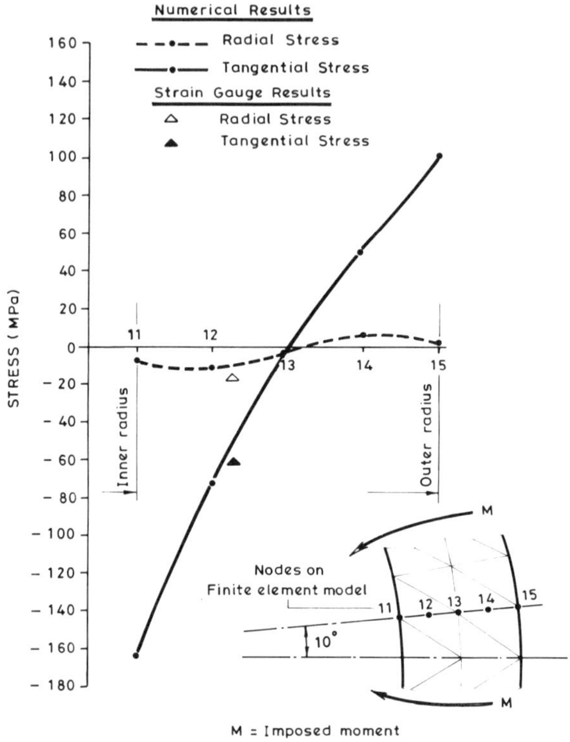

FIG. 3—*Predicted and measured stresses in test ring.*

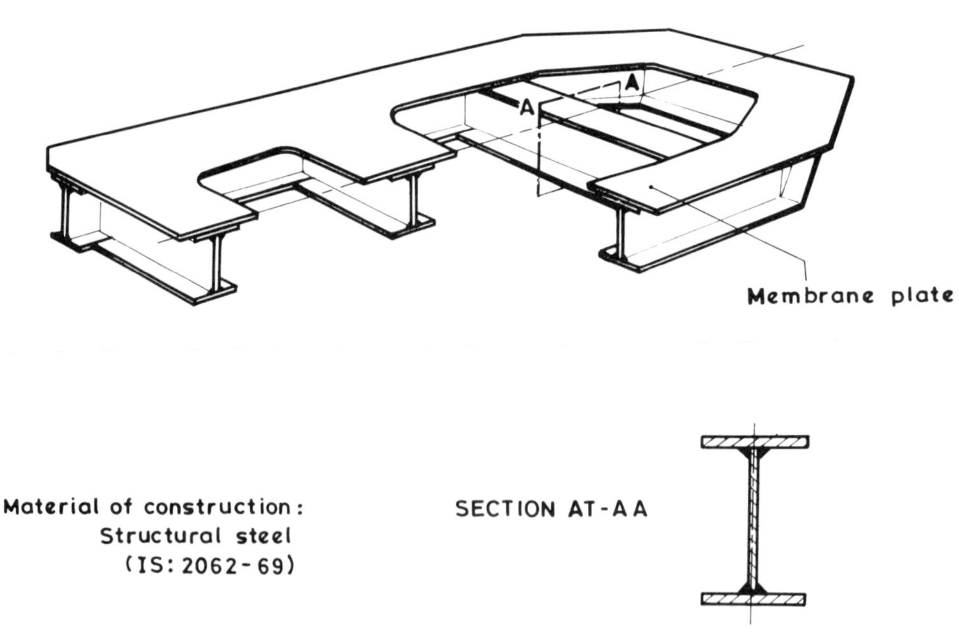

FIG. 4—*Bed plate structure with fabricated I-beams.*

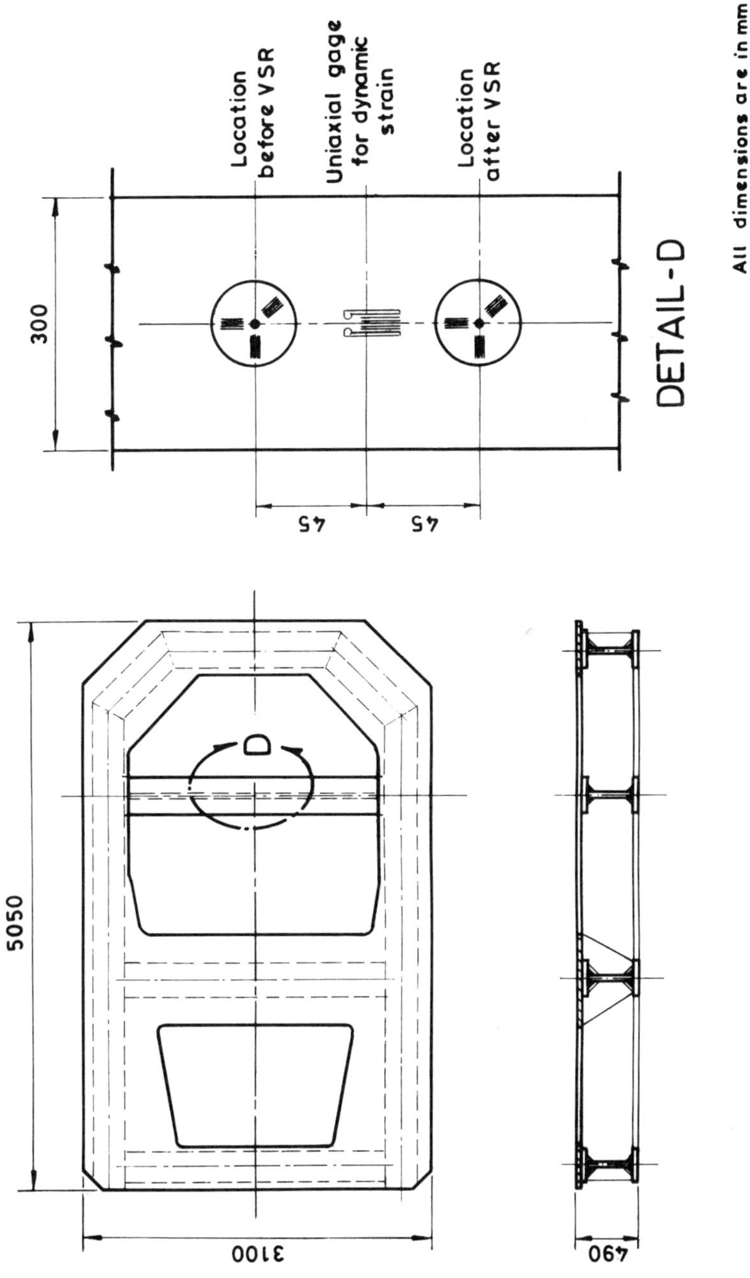

FIG. 5—*Location of strain gages on bed plate.*

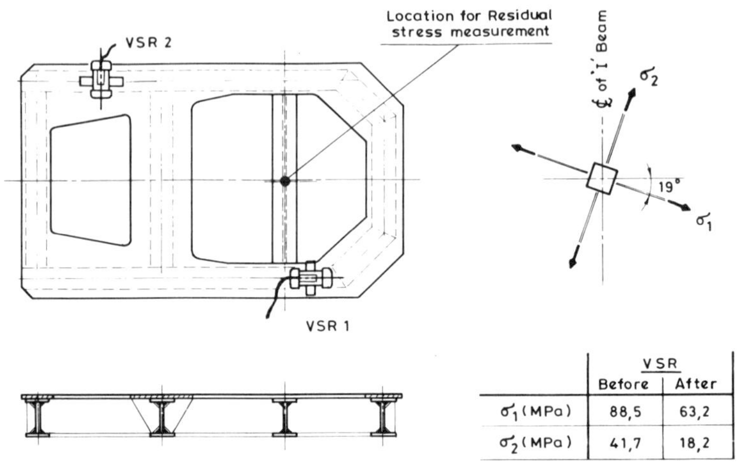

FIG. 6—*Residual stress in bed plate.*

Figure 5 is a schematic of the bedplate structure showing the location of the two residual stress gages and the one uniaxial gage used in the test. Series EA-06-062RE-120 residual stress gages (Photolastic Inc., USA) were used. The uniaxial gage had a 10 mm (25/64 in.) gage length. Previous experience making residual stress measurements in regions of compressive residual stresses enabled us to locate strain gages so as to be in a tensile stress region. The cross member not covered by the membrane plate was selected for the location of the gages. The discussion on residual stresses in T-joints [2] gave us confidence that the stresses at the selected location would be tensile.

Commercially available vibratory equipment was used for the treatment. The vibrator is an a-c motor with masses placed eccentrically on its shaft. A certain directionality of the vibratory force exists in relation to the axis of the vibrator motor. Therefore the structure was treated by placing the vibrator as shown in Fig. 6. The orientation of the vibrator at location 1 was expected to excite torsional modes of vibration. The orientation at location 2 was expected to excite the bending modes of vibration of the bedplate. The structure was supported on rubber pads.

During treatment at location 1, the vibrator speed was varied continuously from rest to

TABLE 1—*Test results.*

Parameter	Time of Measurement	
	Before Treatment	After Treatment
Strain change, (microstrain):		
$\epsilon 1$	−82	−66
$\epsilon 2$	−74	−59
$\epsilon 3$	−10	5
Hole radius, mm	0.84	0.86
$r = (D/D_0)$	3.06	2.99
Residual stresses, MPA:		
$\sigma 1$	88.5	63.2
$\sigma 2$	41.7	18.2
Angle, beta (degrees)	19.0	19.6

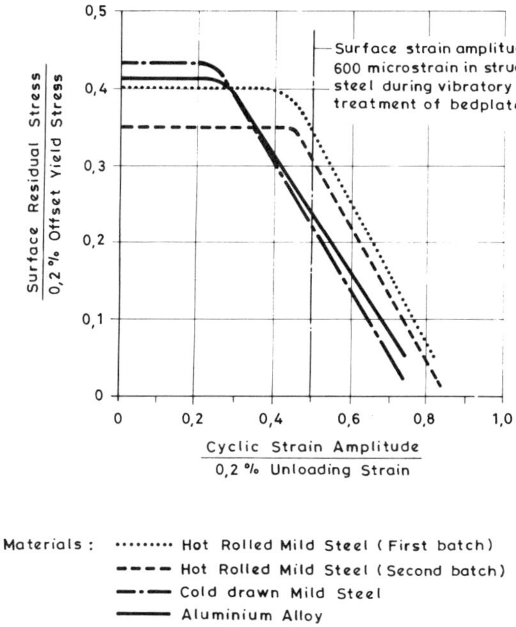

FIG. 7—*Non-dimensionalized residual stress reduction curves.*

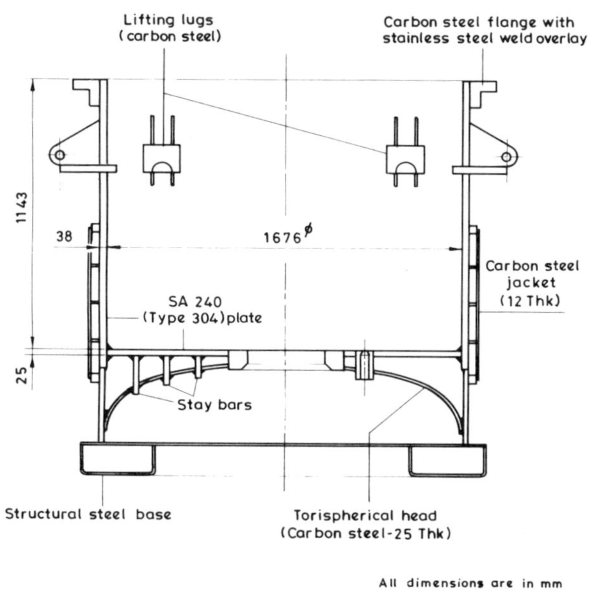

FIG. 8—*Bowl assembly with composite metal construction.*

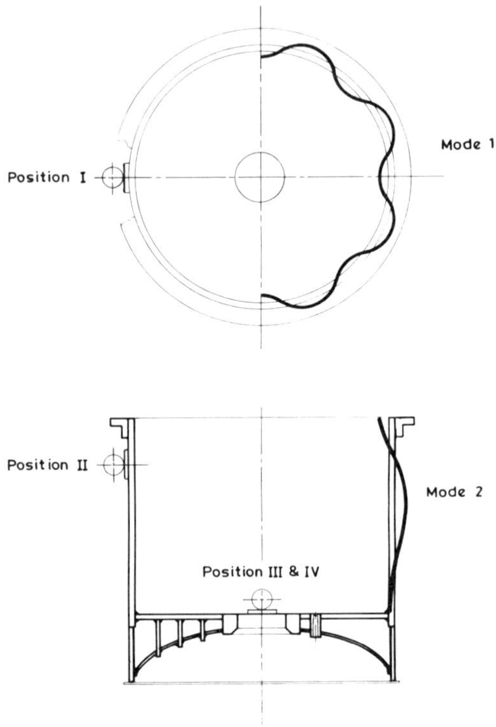

FIG. 9—*Location of vibrator on bowl assembly.*

200 Hz. It was found (from sawdust patterns on the structure) that the vibrator was sited close to an antinode location of the vibrating bedplate. The vibrator was shifted nearer to a nodal area.

An accelerometer mounted on the bedplate allowed the resonance frequencies to be determined. With the vibrator at location 1, the predominant frequencies were observed to be 90 Hz and 144 Hz. At each of these frequencies the vibrations were maintained for about 30 s. At location 2 the frequencies for extended vibration were found to be 112 Hz and 164 Hz.

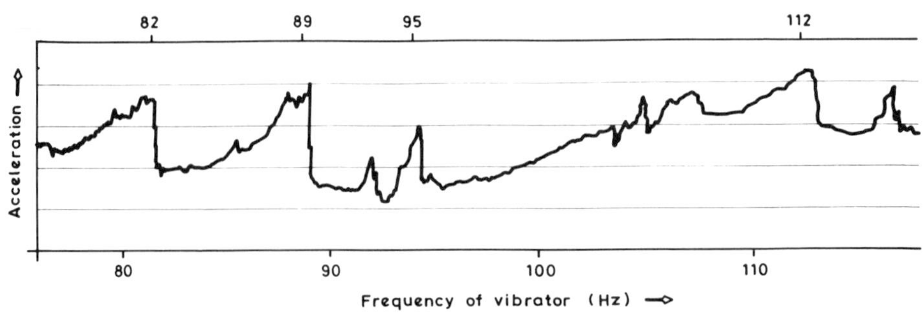

FIG. 10—*Typical acceleration spectrum during vibratory treatment of bowl assembly.*

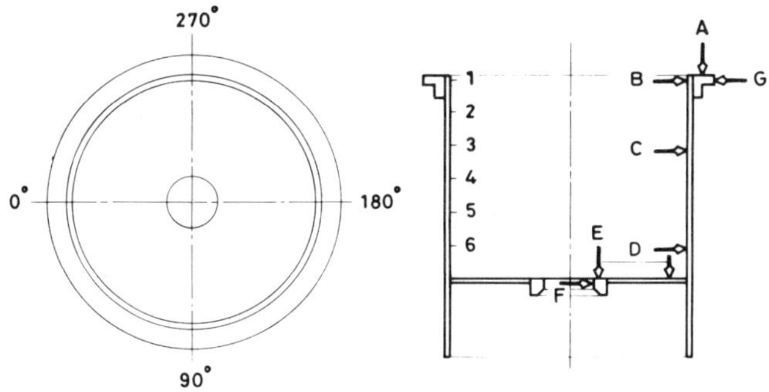

INTERNAL DIAMETER (mm)

Station	Maximum	Minimum
1	1676,425	1676,400
2	1676,500	1676,450
3	1676,550	1676,515
4	1676,500	1676,475
5	1676,450	1676,425
6	1676,465	1676,440

For all stations :

Maximum I.D. = 1676,550 mm
Minimum I.D. = 1676,400 mm
Required I.D. = 1676 $^{+0,40}_{+0,65}$ mm

RUN-OUT AT VARIOUS LOCATIONS (mm)

Location	A	B	C	D	E	F	G
Variation in dial gauge reading	0,01	0,01	0,03	0,03	0,05	0,01	0,01

Height = 1143 $^{+0,025}_{-0,000}$ mm

FIG. 11—*Dimensions of bowl assembly after final machining.*

Results

Residual stress measurements were carried out before and after the vibratory treatment. A 1.59-mm (1/16-in.) diameter end mill was used for drilling of 1.59-mm (1/16-in.) deep blind hole. The end mill was driven at slow speed by an electric drill so as to minimize heating at the gage location. The results are given in Table 1.

The results are shown in Fig. 6. It is observed that the principal stress directions before and after treatment were about the same. As expected from the residual stress pattern in fillet-welded T-joints [2], the principal stresses $\sigma 1$, $\sigma 2$, are tensile and are oriented approximately normal and parallel, respectively, to the axis of the beam. The stress normal to the axis is greater than $\sigma 2$, and reflects the bending effect on the flange due to shrinkage in the fillet-welds.

The residual stress $\sigma 1$ reduced by about 30 % while $\sigma 2$, smaller than $\sigma 1$, reduced by about 57 %. Although a detailed dynamic analysis of the bedplate has not been carried out, it is evident that the beam on which the stresses are measured mainly experiences bending along its longitudinal direction, thus aiding the reduction of $\sigma 2$. Reduction in $\sigma 1$ can occur only if large in-plane vibration of the beam flange occurs; the overall stress relief appears to be due to redistribution following stress relief in the significant direction. The bedplate was subsequently machined, and showed good stability of dimensions.

During resonance, the uniaxial gage indicated dynamic strain amplitudes of about 600 microstrain. As suggested by Dawson [9], the surface strain amplitude is of critical importance in determining whether surface residual stresses would be relieved (Fig. 7). For the steel used, the ratio of surface strain amplitude to the yield strain is about 0.5. As seen from Fig. 7, a reduction in surface residual stresses at the location of measurement is indicated.

Vibratory Treatment of a Bowl Assembly

The bowl assembly structure fabricated at our works featured a jacketed construction (Fig. 8). The bowl has a circular cross-section. The shell and bottom of the bowl were constructed from austenitic stainless steel (Type 304) plate. The shell had several attachments of carbon steel, namely, a heavy flange and lifting lugs, and a cooling water jacket. The bottom of the bowl was stiffened by staying it with carbon steel bars welded to it and to a torispherical carbon steel head. Since an agitator rotates in the bowl, the bowl was required to retain close tolerances on circularity. The composite construction precluded thermal relieving of stresses prior to machining.

It was decided to treat the structure by attaching a vibrator first to the shell and then to the base. Visualization of idealized mode shapes of vibration of the shell indicated that the vibrator should be oriented in two positions as shown in Fig. 9.

Welding of the carbon steel flange and lifting lugs was achieved with the shell stayed by means of spiders welded to its internal diameter. The structure was vibrated twice: before and after detaching the spiders. The vibrations were maintained for 30 s at each of the natural frequencies of the bowl as determined from plots of vibration level (acceleration) versus vibrator frequency (Fig. 10).

Subsequent to the two sets of vibratory treatment, the assembly was machined to final dimensions. Measurements of the diameter taken after final machining remained constant with time and satisfied the requirement that the ovality be within 0.25 mm on a diameter of 1.7 m (see Fig. 11). The run-out figures indicate the change in readings of a fixed-dial gage when the bowl assembly is rotated about its axis through one turn. Two such bowl units are installed in the plant and are operating satisfactorily.

Conclusions

This paper presents work carried out at Larsen and Toubro Limited and establishes the effectiveness of the vibratory stress relieving technique. The advantages of the technique were not appreciated in India until a few years ago. Although a considerable amount of literature is available about the use of vibratory conditioning, the stress relief of structural components has rarely been quantified.

The strain-gage based blind-hole drilling technique was used to measure residual stresses in a fabricated plate structure before and after a vibratory treatment; residual stress relief of about 30 to 57 % was measured at one location on the structure. Since the strain amplitude during resonance is not uniform over the structure, differing levels of residual stresses will remain throughout the bedplate. It is likely that the stress relief would have been greatest at regions of stress concentration, and at surfaces where the highest strain amplitudes are imposed due to vibration of the plate structure in bending modes. The redistribution of residual stresses (as demonstrated by Dawson [9] in cantilever beam specimens) that occurs due to vibration appears to be useful when dimensional stability is important.

Surface strain amplitude at the same location during resonance was measured to be about 600 microstrain. The strain amplitude level appears to have been large enough to cause some stress relief. Such a measurement allows the engineer to determine whether stress relief will occur at the location of interest and to select vibrator frequency and force so that the required strain amplitude is imposed on the structure. Although stress relief by vibrations depends on the material properties and heat treatment, strain amplitude could serve as a measurable physical quantity analogous to temperature in thermal stress relieving. The work of Dawson [*9*] and Soto Raga [*10*] needs to be extended to structures with multiaxial residual stress distributions, and the stress relief that may be obtained due to the surface strain amplitudes imposed during resonance of arbitrarily shaped structures.

At Larsen and Toubro Limited, vibratory conditioning was applied to several types of fabricated structures. In all cases, dimensional stability was maintained and the components continue to operate satisfactorily in service.

Residual stress measurements using the strain-gage based methods are expensive and tedious to perform; stability of dimensions should rather be used as proof of effectiveness of the vibratory conditioning treatment. Observations during the vibratory treatment of fabricated bowl assemblies were presented in this context.

References

[*1*] Masubuchi, K., *Analysis of Welded Structures,* Pergamon, 1980.

[*2*] *Cast Metals Handbook,* American Foundrymen's Society, 1957.

[*3*] Caubo, M., "Determination of the Conditions of Stress Relieving of Welded Assembly," *Welding Journal,* Vol. 42, No. 6, 1963, pp. 282s–288s.

[*4*] Adoyan, G. A., *et al.,* "The Vibratory Stress Relieving of Castings," *Machines and Tooling,* Vol. 38, No. 8, 1967, pp. 18–22.

[*5*] Claxton, R. A., "Vibratory Stress Relieving—Part I: Theory," *Verifact,* Vol. 10, No. 1, February 1983.

[*6*] Weiss, S., Baker, G. S., and Dasgupto, R. D., "Vibrational Residual Stress Relief in a Plain Carbon Steel Weldment," *Welding Journal,* February 1976, pp. 47s–54s.

[*7*] Wozney, G. P. and Crawmer, G. R., "An Investigation of Vibrational Stress Relief in Steel," *Welding Journal,* Vol. 47, No. 9, September 1968, pp. 411s–418s.

[*8*] Zubchenko, O. I., Gruzd, A. A., Torekhov, G., and Sostin, A. G., "Vibrating Loads used for Relieving the Residual Stresses in Welded Frames," *Automatic Welding,* Vol. 27, No. 9, 1974, pp. 59–62.

[*9*] Dawson, R., "Residual Stress Relief by Vibration," Ph.D. Thesis, Liverpool University, December 1975.

[*10*] Soto Raga, A. R., "An Analysis of the Mechanism of Reduction of Residual Stresses by Vibration," PhD Thesis, Georgia Institute of Technology, April 1983.

[*11*] "Measurement of Residual Stresses by the Blind Hole Drilling Method," Technical Data Bulletin T-403, Photolastic Inc., Measurements Group, Wendell, NC, USA.

C. Bouhelier, P. Barbarin, J. P. Deville, and B. Miege[1]

Vibratory Stress Relief of Welded Parts

REFERENCE: Bouhelier, C., Barbarin, P., Deville, J. P., and Miege, B., **"Vibratory Stress Relief of Welded Parts,"** *Mechanical Relaxation of Residual Stresses, ASTM STP 993,* L. Mordfin, Ed., American Society for Testing and Materials, Philadelphia, 1988, 58–71.

ABSTRACT: What is the efficiency of residual stress relieving by mechanical vibration? In this paper, studies were carried out on two primary welded reducer gear boxes weighing 1200 kg each, which were thermally (A box) and mechanically (B box) stress relieved. X-ray stress measurements were made in both cases before and after relieving. Dimensional measurements were made on the B box. During the mechanical treatment the dynamic behavior of the box was followed by acceleration measurements and extensometric testings, in order to estimate the magnitude of the vibration-induced stresses and to detect the eventual presence of a new permanent strain. In the case of thermal relieving, the reduction of residual stresses is about 70 to 100 %. In the case of mechanical relieving, the reduction is between 45 to 100 % for tensile stresses, whereas it is only 0 to 45 % for compressive stresses. Some tests were performed on welded parts of a pump unit component, and the results are similar to the preceding.

KEY WORDS: vibratory stress relief, carbon steel, welded part, stress relief, X-ray stress measurement, residual stress

Tests on Reducer Gear Box Casing

Definition of the Tests

The tests were carried out on two primary reducer gear box casings weighing 1200 kg each which were of normal construction and made of E-26-4 grade steel, and manufactured by Brissoneau-Lotz Marine (Fig. 1). Two types of stress relief were performed: stress relief by heat treatment on casing A, and vibratory stress relief on casing B.

Stress relief of welded parts usually is achieved using heat treatment, so measurements on casing A serve as a reference. The efficiency of the stress relief was checked by X-ray stress measurements, both before and after the stress relief process. During the vibratory stress relief process, measurements of acceleration were made to determine the dynamic behavior of the structure. The existence of a resonance frequency was checked and the deformation of the part was described.

Extensometric measurements were made to define the extent of stresses induced during the course of the vibratory process, and to detect any new permanent deformation after the process was completed. Measurements both before and after stress relief made it possible to keep a check on the changes of the overall dimensions and surface flatness of the parts. These different measurements are given in Table 1.

Vibratory Stress Relief Operating Parameters

Figure 2 shows the part during the vibratory stress relief process using the Stress Relief Engineering Company's "formula 62" apparatus. This machine consists of a variable rotation

[1] Engineers, Centre Technique des Industries Mécaniques (CETIM), Senlis, France 60300.

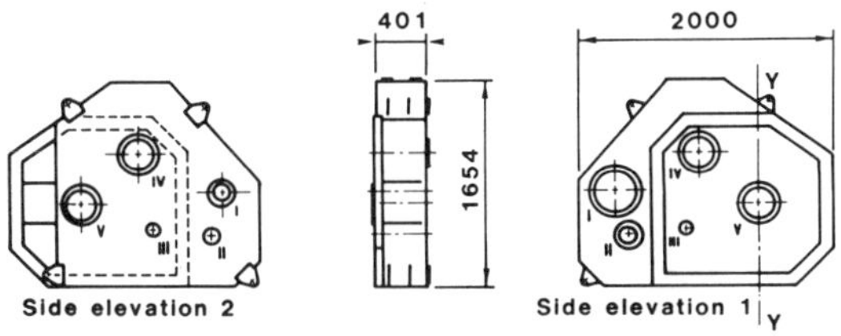

FIG. 1—*Primary reducer gear box casing of pattern B. The pattern A casing is symmetrical with B in relation to axis YY.*

speed, unbalanced mass motor attached to the part by means of clamps. An accelerometer attached to the part makes it theoretically possible to detect the primary frequency of the part. The mechanically welded part is placed on four rubber shock-absorbers.

Vibratory and extensometric analysis shows that the stimulating frequency used in this first stress relief process is not the same as the structure's natural frequency. It corresponds to an overall movement of the casing on the test suspension. Also, the stimulating frequency indicated by the unbalanced mass motor control unit is not the same as the real stimulating frequency measured by CETIM's accelerometers (50 Hz indicated for 30 Hz in real terms). An additional stress relief process at a frequency of 44 Hz, corresponding to the first natural frequency determined by the impact response of the part, was therefore carried out.

Some research indicates that stress relief only occurs for significant deformations, requiring that the part be made to vibrate at a level equal to its natural resonance level [*1–3*]. A more recent publication indicates that better efficiency can be obtained using a subresonant vibration process [*4*].

Table 2 lists the conditions under which vibratory stress relief was performed.

Heat Treatment Parameters—Heat treatment conditions were as follows:

- Temperature rise time: 3 h.
- Held at 620°C for 1 h.
- Cooled down in the furnace.

TABLE 1—*Definition of the Tests.*

Measurements	Casing Mark	Metrological Check	Extensometric Check and Vibratory Analysis	X-ray Stress Measurements
Before stress relief	A			X
	B	X		X
During vibratory stress relief	B		X	
After vibratory stress relief	B	X		X
After stress relief by heat treatment	A			X
One year after first vibratory stress relief process	B	X		

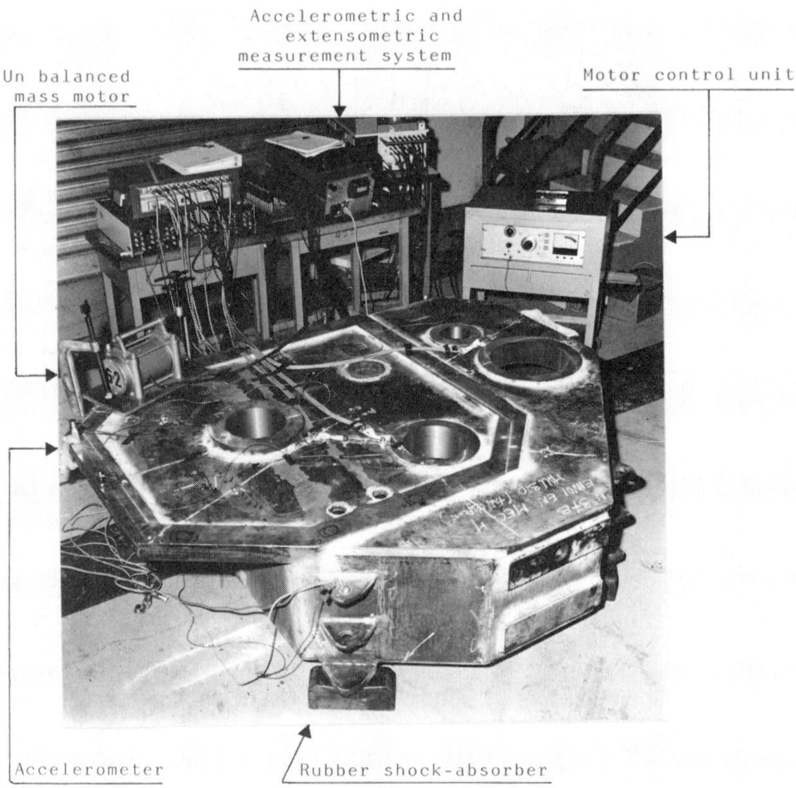

FIG. 2—*Casing during vibratory stress relief.*

X-ray Stress Measurement Operating Parameters—X-ray stress measurements are based on material lattice spacing variations in measurements in several ϕ, ψ directions [*5,6*]. A portable apparatus composed of our laboratory's fittings and with the same characteristics as a conventional diffractometer, has been in service at CETIM for several years (Fig. 4). It was used for all the measurements presented here (Fig. 5).

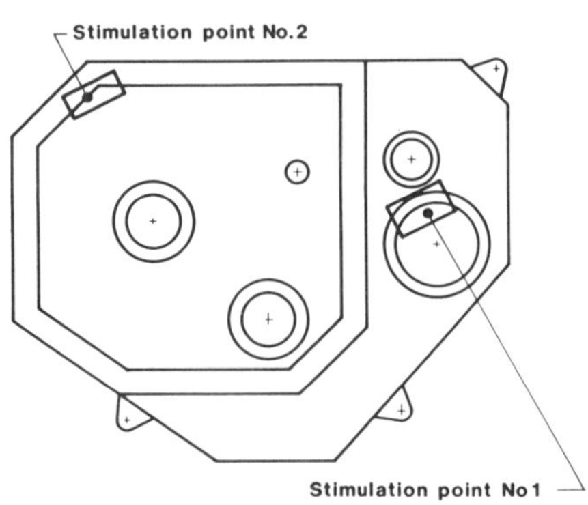

FIG. 3—*Position of stimulation points.*

TABLE 2—*Vibratory Stress Relief Conditions.*

Stress Relief Process	Frequency Indicated by "Formula 62" Apparatus, Hz	Frequency Measured by CETIM, Hz	Vibration Time, min	Position of Unbalanced Mass Motor (Fig. 3)
first	50	30	15	position 1
first	49.5	30	15	position 2
second	...	44	10	position 1
second	automatic	33.6	5	position 1
	cycle	51.8	5	
		39.8	5	

X-ray chromium $\kappa\alpha$ radiation diffracted by the (211) planes from the steel is collected on a position-sensitive detector for several ψ exposures in any ϕ direction. Diffraction profiles are located by a new centroid method [7] and stresses are obtained by general linear regression. In the present case the relationship $2\theta\phi\psi$ versus $\sin^2 \psi$ was very linear, so we had only to consider a biaxial state.

Seven ψ exposures were carried out in 3ϕ directions. In such conditions the accuracy of the measurements essentially depends upon the statistical error and gave standard deviations of about ± 10 MPa (95%).

Results

Metrological Check—Checks of dimensions before and after the first and second vibratory stress relief processes show variations of a magnitude within the accuracy of the measurements (Table 3, Fig. 6a). No significant divergence in surface flatness was recorded (Table 4, Fig. 6b). Vibratory stress relief does not cause any significant variation of the overall dimensions or the surface flatness of the part. The efficiency of stress relief cannot, therefore, be checked by metrological measurements.

FIG. 4—*X-ray stress measurement device developed at CETIM.*

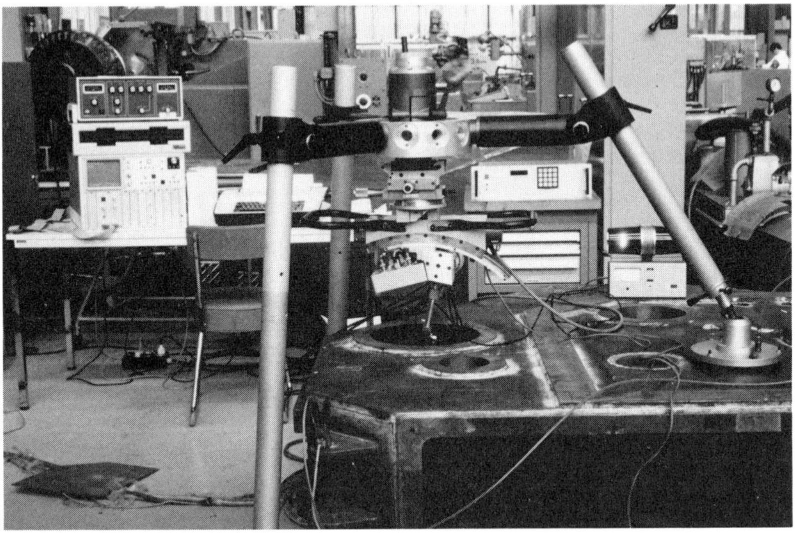

FIG. 5—*X-ray stress measurements on a reducer gear box.*

Measurement of Residual Stresses—The measurement point locations are defined in Figs. 7, 8, 9, and 10. The reduction of the residual stresses by heat treatment varied from 70 to 100 % (Figs. 11 and 12). After relieving, the maximum residual stresses are near ±30 MPa. These results confirm the efficiency of the heat treatment cycle. The reduction of the residual stresses, after the first vibration treatment, varied from 45 to 100 % for tensile stresses, and from 0 to 45 % for compressive stresses (Figs. 13 and 14). After the first stress relief, the maximum residual stresses were −200 MPa in compression and +60 MPa in tension. The second vibration treatment did not cause any further detectable change to the residual stresses (Figs. 13 and 14).

Vibratory and Extensometric Analysis—The first vibratory treatment was carried out at a frequency of 30 Hz, which does not correspond to any natural frequency of the part. The

TABLE 3—*Check of dimensions before and after stress relief (measurement accuracy is equal to ±0.02 mm/m) (Fig. 6a).*

Reference Point Measured	Reading Taken Before Stress Relief, mm	Readings Taken After Stress Relief, mm	
		First	Second
Length: A	1998.42	1998.36	1998.35
Width: B	1640.12	1640.13	1640.10
Center distance of axes			
C	897.64	897.65	897.65
D	697.82	697.84	697.82
E	399.44	399.44	399.46

TABLE 4—*Surface flatness divergences before and after stress relief (point measurement uncertainty: 0.05 mm) (Fig. 6b).*

Reference Point Measured	Divergences Recorded Before Stress Relief, mm	Divergences Recorded After Stress Relief, mm	
		First	Second
1	0	0	0
2	−0.02	+0.02	0
3	0	+0.01	0
4	+0.02	+0.01	0
5	+0.03	0	0
6	0	0	0
7	+0.01	−0.03	−0.01
8	0	0	0
9	−0.01	−0.03	−0.04
10	0	−0.02	−0.04
11	−0.02	−0.01	−0.02
	Divergence of ◇ = 0.05 mm (before)	Divergence of ◇ = 0.05 mm (after)	◇ = 0.04

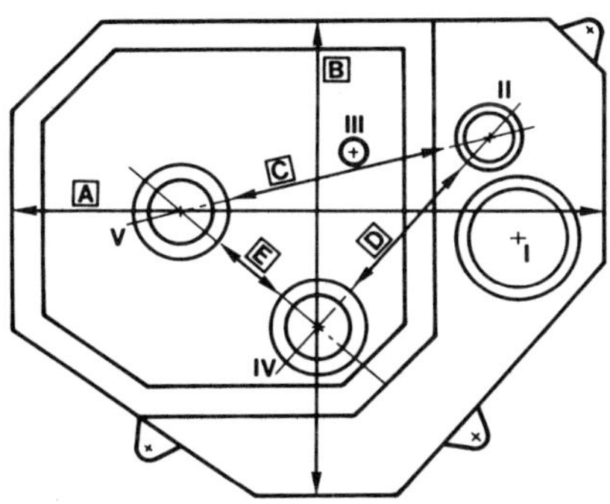

FIG. 6a—*Pattern B casing, check of dimensions before and after vibratory stress relief.*

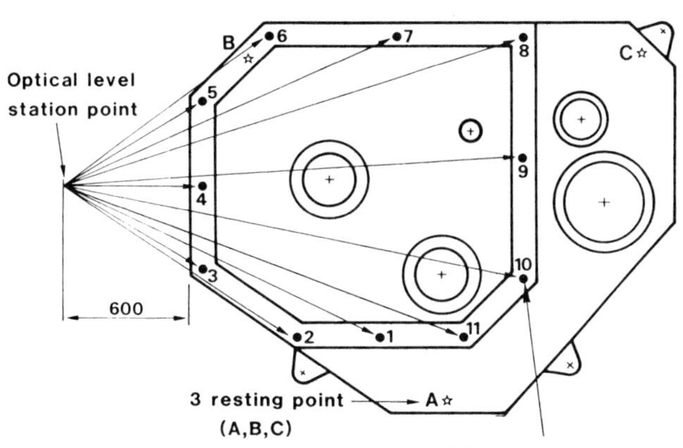

FIG. 6b—*Pattern B casing, surface flatness divergences before and after vibratory stress relief.*

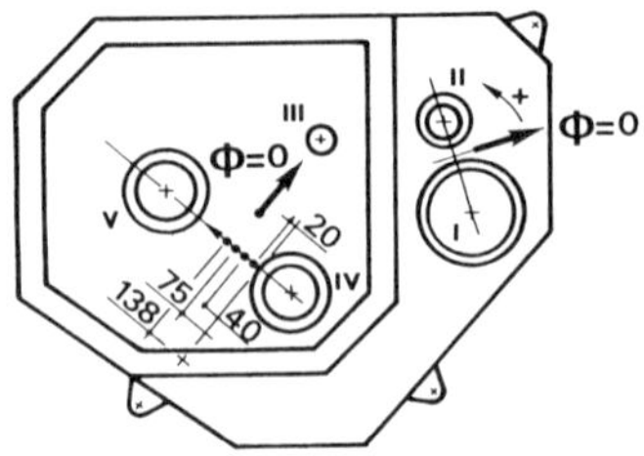

•••• Locations of measurement point of residual stresses
⟶ Direction of angle Φ=0

FIG. 7—*Pattern B casing, side 1 (for vibratory stress relief).*

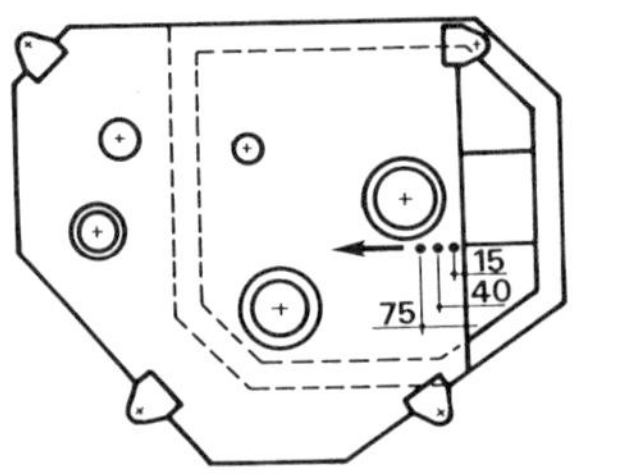

•••• Locations of measurement point of residual stresses
⟶ Direction of angle Φ=0

FIG. 8—*Pattern B casing, side 2 (for vibratory stress relief).*

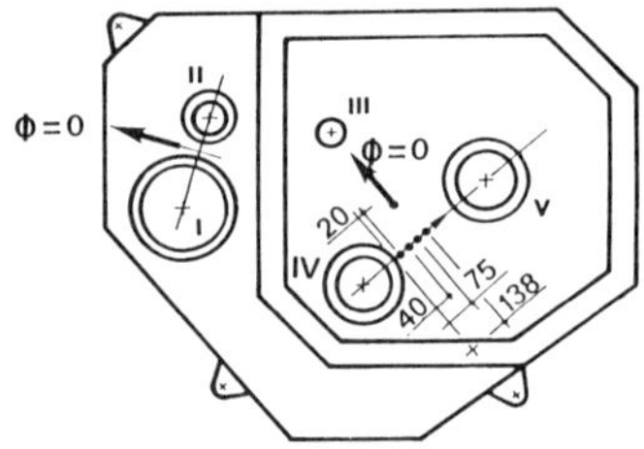

•••• Locations of measurement point of residual stresses
⟶ Direction of angle Φ=0

FIG. 9—*Pattern A casing, side 1 (for heat treatment stress relief).*

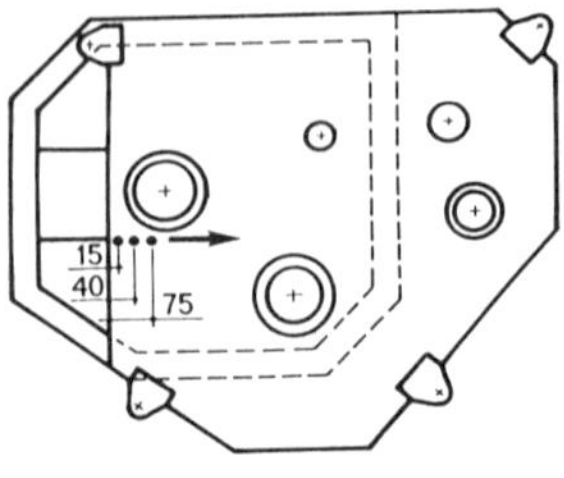

•••• Locations of measurement point of residual stresses
⟶ Direction of angle Φ=0

FIG. 10—*Pattern A casing, side 2 (for heat treatment stress relief).*

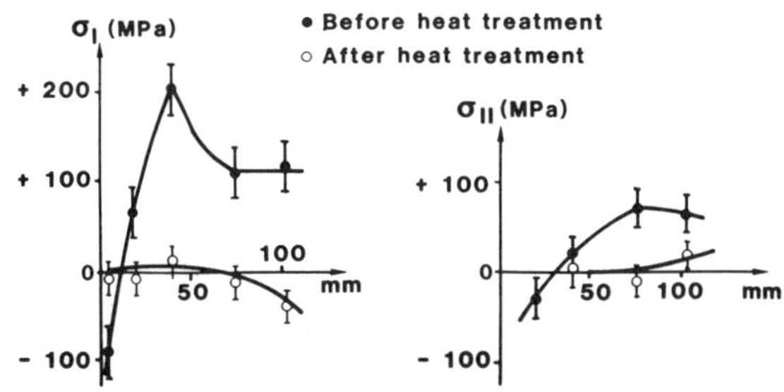

FIG. 11—*Pattern A casing, side 1, variation of the principal stresses* σ_1 *and* σ_{11} *measured along axes IV and V.*

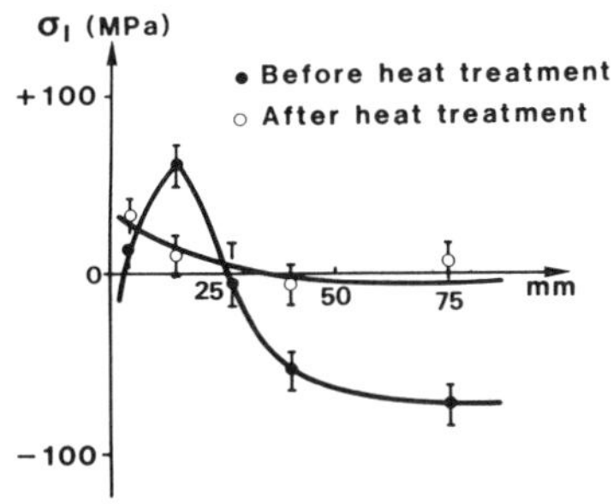

FIG. 12—*Pattern A casing, side 2, variation of the principal stress* σ_1 *(measurement points as in Fig. 10).*

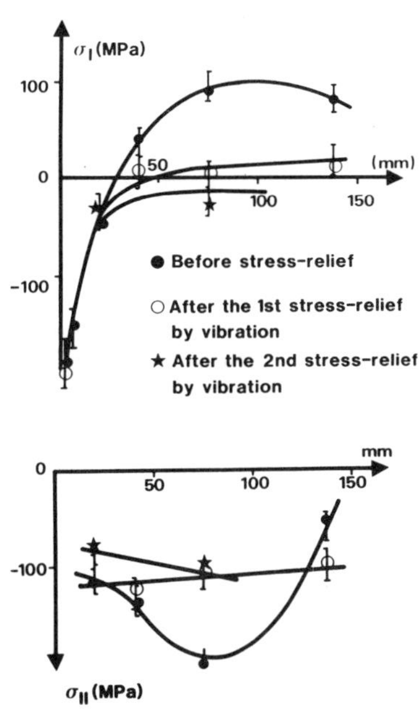

FIG. 13—*Pattern B casing, side 1, variation of the principal stresses* σ_1 *and* σ_{11} *measured along axes IV and V.*

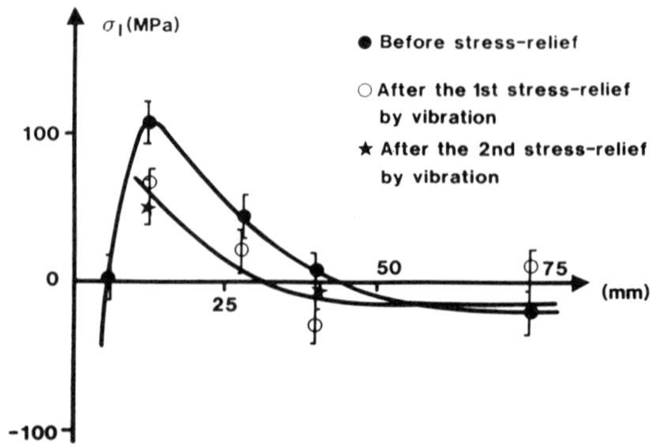

FIG. 14—*Pattern B casing, side 2, variation of the principal stress* σ_1 (*measurement points as in Fig. 8*).

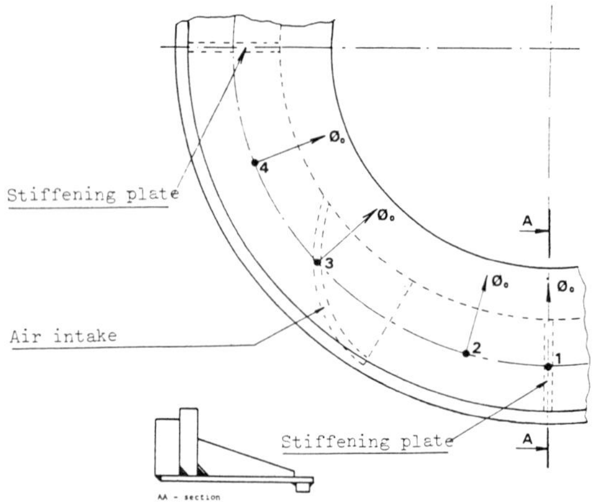

FIG. 15—*Pump unit component, locations of measurement points of residual stresses.*

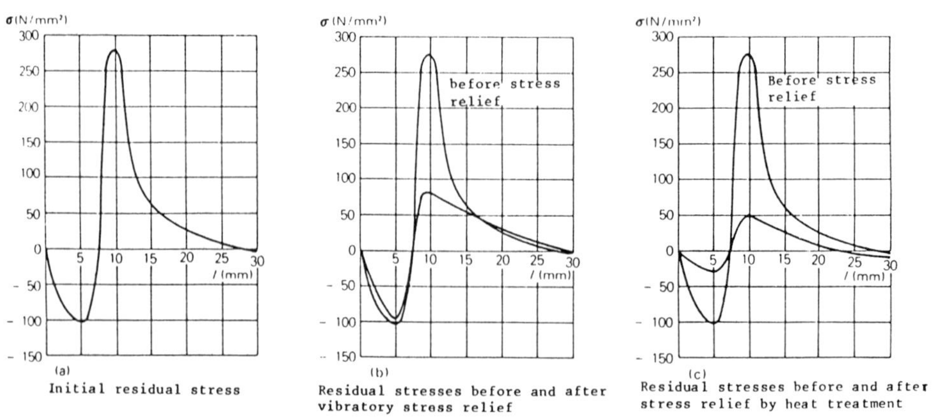

FIG. 16—*Example of stress relief* (*after* [8]). *Unalloyed steel sheet 25 × 150 × 500 mm. Weld in longitudinal direction on part surface. Residual stresses are of order of size of elasticity limit.*

TABLE 5—*Residual stresses before and after vibratory stress relief on two pump unit components.*

		Before Stress Relief			After Vibratory Stress Relief					
Piece Reference	Measurement Point	σ_I, MPa	σ_{II}, MPa	ϕ_{σ_I}/ϕ_o, deg	σ_I, MPa	σ_{II}, MPa	ϕ_{σ_I}/ϕ_o, deg	$\Delta\sigma_I$, MPa	$\Delta\sigma_{II}$, MPa	$\Delta\phi$, deg
X1	1	56	99	−32	33	60	−28	−23	−39	+4
X1	2	−59	55	5	−83	30	−9	−24	−25	−14
X1	3	−49	−16	−9	−52	−21	−5	−3	−5	+14
X1	4	−12	84	2	−25	85	−10	−13	1	−12
V1	1	146	93	2	106	48	21	−40	−45	0
V1	2	65	114	−9	1	76	−26	−64	−38	−17
V1	3	132	244	−8	90	191	−11	−42	−53	−3
V1	4	35	75	−3	14	57	−11	−21	−18	−8

σ_I = principal stresses;
σ_{II} = principal stresses; and
ϕ_{σ_I}/ϕ_o = angle between σ_I and reference direction ϕ_o.

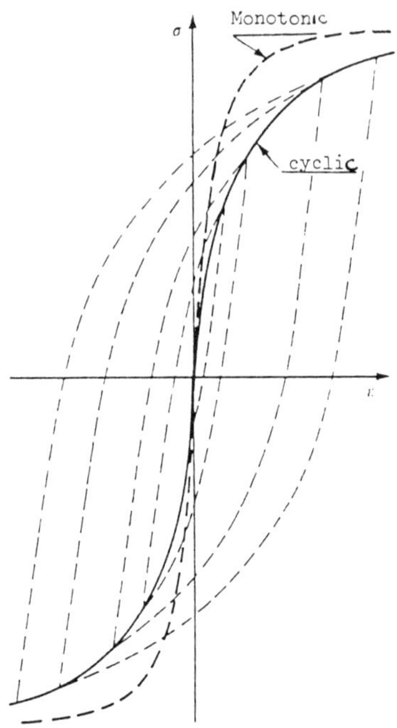

FIG. 17—*Plot of cyclic curve (in solid line) and of the monotonic curve (dotted).*

dynamic stresses induced by vibration and measured by three strain gages placed in the zone of the residual stresses measurement (Fig. 7 side 1: along axes I and II, along axes IV and V) are, at the very most, 1.5 MPa. The second vibratory treatment was carried out at the first natural frequency of the structure (44 Hz). A dynamic stress level of 100 to 200 times that obtained in the first stress relief process was recorded.

Tests on Pump Unit Component

Some tests were carried out on welded parts of a pump unit component made of E24 grade steel (external diameter = 1400 mm, internal diameter = 900 mm, weight = 270

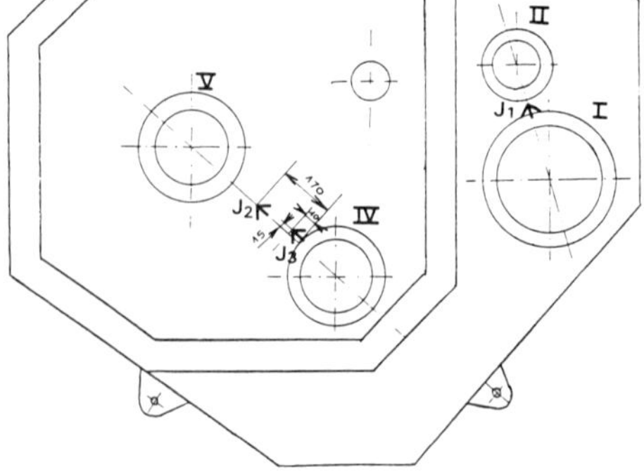

FIG. 18—*Positions of strain gages.*

kg). A vibratory stress relief was achieved with the same apparatus as described before. During this treatment the part was vertically hung with slings, and the vibratory conditions were not recorded.

As before, the efficiency of the stress relief was checked by X-ray stress measurements before and after the stress relief process. Locations of measurement points of residual stresses can be seen on Fig. 15 and the results of measurements are shown in Table 5.

The values $\Delta\sigma$ show without ambiguity the relative efficiency of this kind of treatment on the welded parts tested. Excepting point 4 on part X1, all tensile stress values decreased with variable ratios. The variations of the principal stress directions were not significant. Points 2, 3, and 4 on part X1 show an increase in compressive residual stresses after vibratory stress relief. We cannot explain this phenomenon, but perhaps the vibratory process induces a redistribution of the residual stresses at the scale of the welded structure.

Interpretation of Results

The vibratory stress relief process does have an effect. Tensile residual stresses are reduced more easily than compressive residual stresses. Other similar results have been obtained by other engineers experimenting with stress relief equipment similar to our own (Fig. 16). The vibration frequency of the structure does not seem to have any influence on the efficiency of stress relief. It is not necessary to relieve stress using the part's frequency of resonance. The extent of the dynamic stresses having caused stress relief is of the order of 1.5 MPa for the reducer gear box casing. This result contradicts the majority of analyses [*1*] and [*2*], which put vibratory stress relief down to localized cyclic plastic deformations.

Knowing that, under a cyclic load, the elastic limit of certain metals is greatly reduced (Fig. 17), dynamic stresses below the static elastic limit can cause localized plastic deformation and bring about the reduction of residual stresses. This type of argument supposes that stresses ($\sigma_R + \sigma_{applied}$) are at least equal to the material's cyclic elastic limit; this cannot explain stress relief achieved for applied dynamic stresses less than or equal to 1.5 MPa. The partial stress relief is perhaps due to movements and reorganization of anomalies at the atomic level (dislodging of dislocations, movement of interstitial atoms, internal friction); this hypothesis remains to be verified.

Having no physical explanation of the phenomenon, it is not advisable to extrapolate the results to other structures and other metals.

Conclusion

Vibratory stress relief can be used for stabilization of dimensions before machining of welded parts. It is necessary to carry out a preliminary test to judge how effective the process is for a given manufacture. If the treatment is sufficient, size variations at the time of machining are negligible. This type of stress relief is already used in industry.

The process may be used for stress relief of parts for which a stress relieving heat treatment cannot be used. However, it should be borne in mind that stress relief is only partial.

It is quite obvious that vibratory stress relief should be ruled out completely as a replacement for heat treatment when the latter is applied for metallurgical reasons or when heat treatment is compulsory when following given manufacturing rules.

References

[*1*] Dawson, R. and Moffat, D. G., "Vibratory Stress Relief: a Fundamental Study of its Effectiveness," *Journal of Engineering Materials and Technology,* Vol. 102, April 1980, pp. 169–176.

[2] Weiss, S., Baker, G. S., and Das Gupta, R. D., "Vibration Residual Stress Relief in a Plain Carbon Steel Weldment," *Welding Research Supplement,* Feb. 1976, pp. 47s–51s.
[3] Zuschenko, O. I., Gruzd, A. A., Orekhow, G. I., and Sostin, A. G., "Vibrating Loads Used for Relieving the Residual Stresses in Welded Frames," *Avtomaticheskaja Svarka,* No. 9, 1974, pp. 64–65 (translation in: *Automatic Welding* (URSS), Vol. 27, No. 9, pp. 59–62).
[4] Hebel, A. G., Jr., "Subresonant Vibrations Relieve Residual Stress," *Metal Progress,* Nov. 1985, pp. 51–55.
[5] Lecroisey, F., Miege, B., and Saint Etienne, A., "La Mesure des Contraintes Résiduelles par Rayons X," *Mémoire Technique du CETIM,* No. 33, March 1978.
[6] Miege, B., Barbarin, P., and Convert, F., "La Mesure des Contraintes par Rayons X," To be published by CETIM.
[7] Convert, F., "Groupement Français pour l'Analyse des Contraintes par Diffractométrie X," ENSAM, Aix en Provence, 1982.
[8] Prohaszka, J., Hidasi, B., and Varda, L., "Vibration, Induced Internal Stress Relief," *Periodica Polytechnica,* Vol. 29, No. 1, 1975, pp. 69–78.

DISCUSSION

A. Fox[2] (*written discussion*)—You showed dimensional changes, very small, as a consequence of vibratory stress relief of the gear box. Have you similar measurements obtained before and after thermal treatment?

C. Bouhelier et al. (*authors' closure*)—We have not measured the dimensional changes after the thermal treatment. Our objective was to determine if the efficiency of the vibratory stress relief can be analysed by the measurement of dimensional changes. On the other hand the problem of variations of the dimensions by the thermal treatment is generally well known.

A. Fox[2] (*written discussion*)—What was the thermal stress relieving treatment time and temperature?

C. Bouhelier et al. (*authors' closure*)—Conditions under which heat treatment was carried out were as follow: the temperature rise time was 3 h, was held at 620°C for 1 h, and cooled down in the furnace.

G. B. Ulrich[3] (*written discussion*)—How does one decide where to apply vibration frequencies on a particular part?

C. Bouhelier et al. (*authors' closure*)—The motor is clamped to the workpiece; this piece is placed on four rubber shock-absorbers or is vertically hung with slings. For the vibratory treatment of little pieces, it is more convenient to clamp these onto a table on which the unbalanced mass motor is also fastened. The most effective treatment conditions were found by applying the vibratory load for different positions of the motor. It is necessary to carry out some preliminary tests to judge how effective the process is for a given manufacturer. The criterion of choice of the motor position are the size variations at the time of machining.

N. W. Hung[4] (*written discussion*)—I am very impressed with the technical data and high-quality slides presented by Mr. Flavenot. However, I suspect that there may be some minor experimental errors in your strain gage experiment.

[2] Fox Consulting and Testing Services, 24 Haddonfield Drive, Parsippany, NJ 07051.
[3] Martin Marietta Energy Systems, Y-12 Plant Bldg. 9110 MS-1, Oak Ridge, TN 37830.
[4] Hewlett Packard, 1400 Fountain Grove Parkway, Santa Rosa, CA 95404.

Mechanical vibration, as my proposed mechanism, facilitates dislocation glides in the vibrated material. It reduces the Peierls stress due to the entangled dislocations and lets the material energy level drop down to a lower state, regardless of whether the residual stress is tensile or compressive. On one slide, however, the data shows only tensile residual stresses being reduced, not the compressive components, after the first and second vibration treatment. Also, if there were only a reduction in tensile residual stress, the material would be in static non-equilibrium and distortion would occur. This fact is again in conflict with your data that there was no measurable distortion of the gear box after vibration. The inconsistency of the experimental data also shows on another slide, in which the initial compressive residual stress become tensile after the vibratory treatment.

I think we will learn more if you can provide the accuracy of your stress measurement technique. Also, I would like to know how you measured/calculated the natural frequency of the gear box and how you picked the location of the vibrating motor and its weight. Thank you.

C. Bouhelier et al. (*authors' closure*)—We agree with you, if there were only reduction in tensile residual stress, a distortion would occur. We think that vibratory process induced a redistribution of the residual stresses at the scale of the welded part and we have not measured the residual stresses everywhere on the piece, maybe in different points the compressive residual stresses have been reduced.

The accuracy of a stress measurement technique is defined in the paragraph "X-ray stress measurement operating parameters". Seven exposures were carried out in 3ϕ directions and in such conditions the accuracy of the measurements, gave a standard deviation of about ± 10 MPa. The natural frequency of 44 Hz is determined by the analysis of the impact response of the welded structure measured by five accelerometers. The position of the vibrating motor is determined by the possibility of clamping on the piece. If it is possible, we clamp the motor between two rubber shock-absorbers or at the antinode position if it has been found.

R. W. Hampton[5] (*written discussion*)—It is not clear where the vibratory stresses were measured. Please identify whether the strain gage location was the same as the areas where the residual stress measurements were made. This is important, since stresses in a vibration mode can vary greatly from one spot to another.

In regard to the reported stresses near the welds, please identify the distance to the welds on the horizontal axes. Also, do the authors have any additional data from other weld regions which show the more usual tensile residual stress condition near weldments?

C. Bouhelier et al. (*authors' closure*)—Extensometric measurements were made by three strain gages placed in the zone of residual stress measurement. The positions of these measurements points are presented on Fig. 18 (gage J1 along axes I and II, gages J2 and J3 along axes IV and V). For all the results of stresses measurement the distances on the horizontal axes (Figs. 11, 12, 13, and 14) are from the weld toe. The exact locations of measurement points are indicated on Figs. 7, 8, 9, and 10.

[5] NASA, Ames Research Center, Mail Stop 213-3, Test Engineering and Analysis Branch, Moffett Field, CA 94035.

J. Graham Wylde[1]

Proposed Investigation of Process for Reducing Residual Welding Stresses and Distortion by Vibration (Abstract Only)

As a result of differential contractions that occur during cooling after welding, high tensile residual stresses are formed in welded joints. These residual stresses can constitute major problems both during fabrication and once a structure enters service. During fabrication they can cause large distortions that can be a major obstacle to assembly and, most significantly, to maintaining tolerances in components which must be finish machined. In addition, residual stresses can produce cracking in weldments often necessitating major reworking. Once a structure enters service residual stresses increase the risk of fracture, can contribute to the propagation of fatigue cracks, and increase the risk of stress corrosion cracking in susceptible combinations of material and environment. Thus, the presence of residual stresses has been cited as a contributory factor in many major industrial failures.

The cost to the U.S. economy of the above factors runs into many millions of dollars annually. Consequently, there is a vital need to investigate means of reducing residual welding stresses. Currently, the most widely used technique in industry is postweld heat treatment. However, there are many limitations on the use of this technique and, consequently, alternative methods are needed.

Previously, vibratory stress relief has been shown to be rather inconsistent. It is now considered that a sub-resonant stress relief may offer considerable advantages over alternative vibratory techniques which are not now considered to operate in the optimum mode. If the technique is shown to be effective, every branch of manufacturing in the U.S. using welding will benefit.

The objectives of the proposed program are:

- To investigate the effectiveness of sub-resonant stress relief in reducing distortions and residual stresses by welding.
- To investigate the effect of applying vibrations during the welding operation.
- To review current code requirements and fabrication practices relating to stress relief.

[1] Manager, Engineering Department Edison Welding Institute, 1100 Kinnear Road, Columbus, Ohio 43212

Stress Relaxation in Fatigue

J. Lu, J. F. Flavenot, and A. Turbat[1]

Prediction of Residual Stress Relaxation During Fatigue

REFERENCE: Lu, J., Flavenot, J. F., and Turbat, A., **"Prediction of Residual Stress Relaxation During Fatigue,"** *Mechanical Relaxation of Residual Stresses, ASTM STP 993,* L. Mordfin, Ed., American Society for Testing and Materials, Philadelphia, 1988, pp. 75–90.

ABSTRACT: This paper presents a model for the prediction of the residual stress distribution during and after fatigue. Finite element software is used for incorporating cyclic plasticity into the calculations. A simplified inelastic analysis was applied. Its essential feature consists of the introduction of a group of internal parameters that characterize local inelastic mechanisms and a group of transformed internal parameters that are linearly linked to the previous ones through a symmetrical non-negative matrix. With this approach, the treatment of the local plastic yield conditions can be made easily from simple elastic analysis. This method was applied on two quenched and tempered alloy steels: shot-peened 35NCD16 (nominal 0.32 to 0.39 C, 0.30 to 0.6 Mn, 0.1 to 0.4 Si, 1.0 to 2 Cr, 3.6 to 4.1 Ni, 0.25 to 0.45 Mo) grade steel and ground 42CD4 grade steel (UNS G41420). The influence of different parameters of fatigue testing are studied, including the number of cycles and applied stress. The computed values of relaxed residual stress distributions were compared with experimental results obtained by X-ray diffraction measurements.

KEY WORDS: fatigue, residual stress, relaxation, finite elements method

Introduction

It is well known that residual stresses are present in many mechanical components. These stresses can be introduced by forming and joining processes and by surface treatments. A compressive stress improves the fatigue behavior of materials, while tensile residual stresses produce an opposite effect [*1–4*].

Residual stresses can be relaxed by the deliberate application of mechanical or thermal energy. In this paper, only relaxation during cyclic loading (fatigue) will be studied. In fact, the relaxation phenomenon depends on a complex interaction of a number of factors, such as the stress amplitude, the number of cycles of the loading, the temperature, the state of the initial residual stress, and the nature, the origin, and mechanical properties of the material.

If we want to take into account the residual stresses in a fatigue calculation, it is very important to consider the relaxation of residual stress [*2*]. However, in fatigue design applications, the relaxation phenomenon often is neglected. In only a few cases are the residual stresses systematically analyzed using measurement of the residual stress state during and after fatigue testing. This is usually a difficult, time-consuming task.

Thus, it is necessary to develop methods of calculation for prediction of the relaxation of residual stress during fatigue. A model of relaxation is presented that relates residual stress distribution in the depth plane to the following:

1. The number of fatigue cycles,
2. The cyclic stress amplitude,

[1] Engineers, Centre Technique des Industries Mécaniques (CETIM), Senlis, France 60300.

3. The ratio $R = \sigma_{min}/\sigma_{max}$,
4. The loading direction, and
5. The cyclic hardening behavior of the material.

The model is implemented with finite element software.

Principle of the Calculation Method

When an elastic-plastic material is subjected to a cyclic uniform uniaxial stress that varies between σ_{min} and σ_{max}, it may respond in different ways, as shown in Fig. 1, where σ is the stress and ϵ is the axial strain. For cyclic plasticity problems, we can use computer codes (based on the finite elements method) that are now very efficient. The numerical approach is very expensive and cannot give a definite answer except for a relatively small number of cycles. A new, simple practical approach is proposed by Zarka and Casier [5] for the inelastic analysis of structures. In the case of cyclic plasticity (for example, the relaxation of the residual stress due to a local cyclic plastic strain), this method cannot give the exact solutions. However, this method can predict the response of a kinematic hardening material (kinematic hardening allows the yield surface to translate without changing shape) to the stabilized cycle (elastic shakedown or plastic shakedown) [6].

The new approach of Zarka and Casier introduces transformed internal parameters into the calculation of the states of stress and the strains. These parameters can be calculated by the following relationship:

$$\hat{\alpha} = \alpha - dev\ \rho = \alpha - s + s^{el} \tag{1}$$

where

$$\alpha = C\epsilon^{P}$$

and

$$C = 2/3\ H$$

where

α = tensorial internal variable connected with kinematic hardening,
C = constant,
H = cyclic hardening modulus of the material,

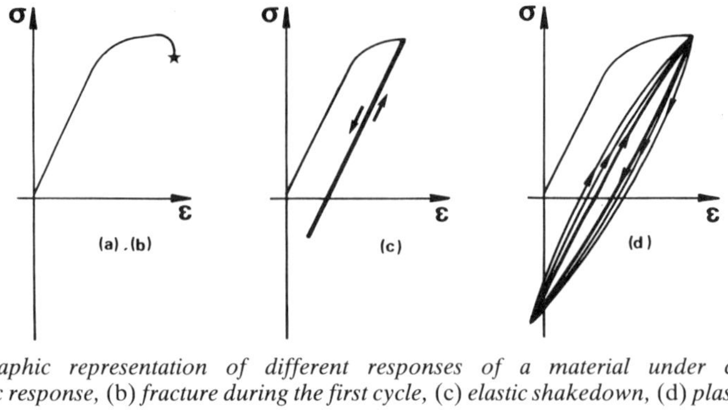

FIG. 1—*Graphic representation of different responses of a material under cyclic loading;* (a) *purely elastic response,* (b) *fracture during the first cycle,* (c) *elastic shakedown,* (d) *plastic shakedown, and* (e) *ratcheting.*

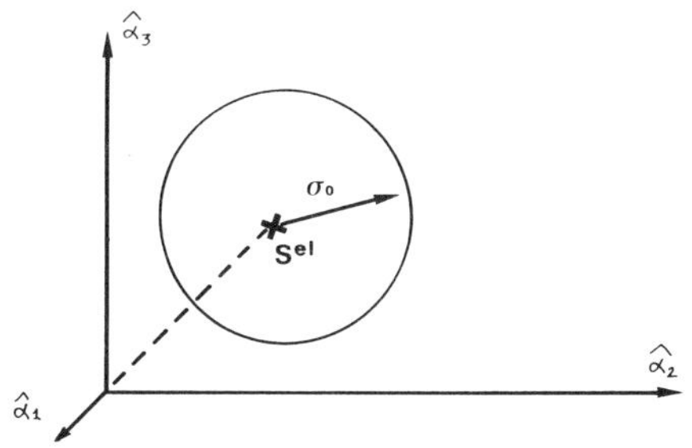

FIG. 2—*Graphic representation of the yield surface in the* $\hat{\alpha}$ *space (similar to a stress space).*

ϵ^P = tensor of the plastic strains,
ρ = tensor of the residual stresses,
S = real stress deviator tensor,
S^{el} = elastic stress deviator tensor,
dev ρ = deviator of residual stress, and
$\hat{\alpha}$ = transformed internal parameters.

For the material following the Von Mises criterion, the yield condition is:

$$\frac{3}{2}(S - \alpha)^T(S - \alpha) \leq \sigma_o^2 \tag{2}$$

where

σ_o = real yield limit, and
$(S - \alpha)^T$ = transposed form of $(S - \alpha)$.

From Eqs 1 and 2 we can deduce the new plastic criterion:

$$\frac{3}{2}(S^{el} - \hat{\alpha})^T (S^{el} - \hat{\alpha}) \leq \sigma_o^2 \tag{3}$$

This relationship defines a new yield surface (or sphere) using the transformed parameter $\hat{\alpha}$. The center of this sphere is S^{el} and its radius is σ_o (Fig. 2).

In the space of the transformed parameters, the interpretation of the plastic criterion is as follows: if the state of nonlinearity (plasticity) of the material is defined by the tensor of the parameters, $\hat{\alpha}$, the representative point of the state of the stress is in the interior of the

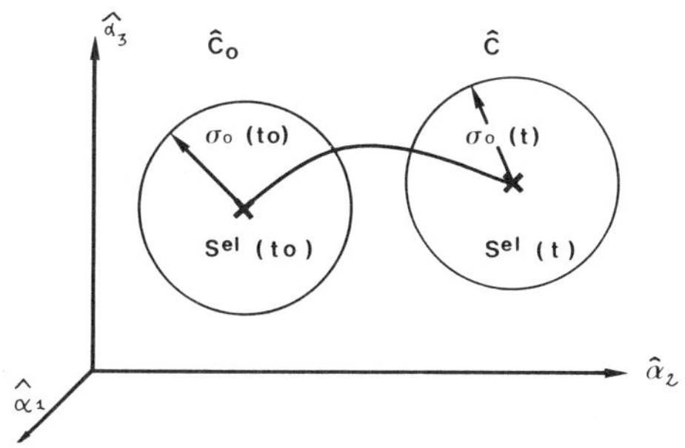

FIG. 3—*Graphic representation of the evolution of the plastic criterion when the loading condition changes between initial time,* t_o, *and time,* t, *(any time).*

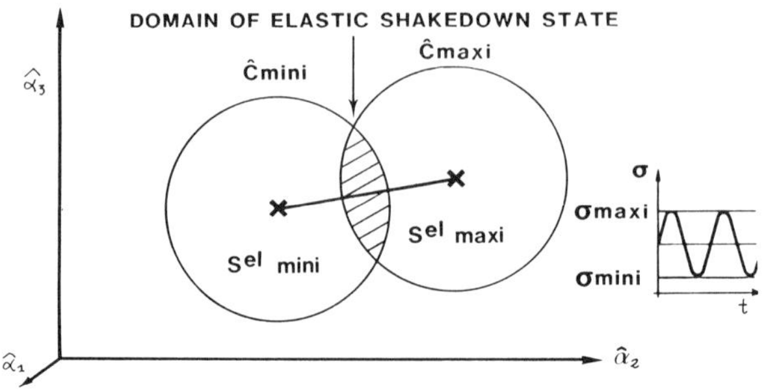

FIG. 4—*Graphic representation of the domain of elastic shakedown state in* $\hat{\alpha}$ *space.*

sphere with the center, S^{el}, and the radius of σ_o at all times.

When the loading condition changes with time, the position, S^{el}, and the size, σ_o, of the sphere also change (Fig. 3), but these changes depend only on the parameters calculated by the elastic calculations.

During a constant amplitude fatigue test ($\sigma_{mini} < \sigma < \sigma_{maxi}$), the representative point of the state of stress evolves between two convex sets, $\hat{C}_{mini}$ ($S^{el}{}_{mini}$, $\sigma_{o\ mini}$) and $\hat{C}_{maxi}$ ($S^{el}{}_{maxi}$, $\sigma_{o\ maxi}$) in the $\hat{\alpha}$ space. The convex set, $\hat{C}$ (σ), is subjected to a simple radial translation between the two limiting positions, $\hat{C}_{mini}$ and $\hat{C}_{maxi}$.

According to the Zarka and Casier criterion, for a kinematic hardening material, the condition for elastic shakedown (stabilization of the residual stress state) is that the convex set $\hat{C} = \hat{C}_{mini} \cap \hat{C}_{maxi}$ is a non-void set.

In the graphic representation of the parameter $\hat{\alpha}$, we need to see if the convex sets $\hat{C}_{mini}$ and $\hat{C}_{maxi}$ have a non-void intersection (Fig. 4).

The calculation of the stabilized state of the residual stress after fatigue loading consists in verifying the condition of the elastic shakedown:

$$\sqrt{\frac{3}{2}\,(\Delta S^{el})^{T}(\Delta S^{el})} \leq 2\sigma_o \tag{4}$$

where

$\Delta S^{el} = S^{el}{}_{maxi} - S^{el}{}_{mini}$, and
$(\Delta S^{el})^{T}$ = transposed form of ΔS^{el}.

Details of the different steps in the calculation of the stabilized stress state in the case of elastic shakedown by an approximate method is found in Ref *5*.

This method is very interesting. Elastic calculations for only a small number of cycles are necessary to find the limiting state (elastic shakedown) using the initial residual stress field, the cyclic hardening characteristics of the material, and the cyclic applied loading. With this approach, a computer program using the finite element method was developed by the CETIM (CA.ST.OR software, option PC2D) for calculating the $\hat{\alpha}$ in all points of the part. This

TABLE 1—*Chemical composition of 35NCD16 steel.*

C	Mn	Si	Ni	Cr	Mo	P, max	S, max
0.355	0.34	0.35	3.7	1.72	0.33	0.009	0.003

TABLE 2—*Mechanical properties of 35NCD16 grade steel.*

Yield Strength, MPa	Ultimate Tensile Strength, MPa	Elongation, %	Reduction in Area, %
1000	1100	17.5	65

software is only applicable in the case of elastic shakedown for a radial cyclic loading. Cyclic loading is said to be radial if we can write:

$$\sigma(t) = \lambda(t)\sigma_{mini} + (1 - \lambda(t))\sigma_{maxi} \tag{5}$$

where

$\lambda(t)$ = periodic function of time varying monotonically between 0 and 1 and then between 1 and 0, and

σ_{mini} and σ_{maxi} = two constant global stress states.

Application of the Simplified Calculation Method for the Prediction of the Relaxation of Residual Stresses During Fatigue

Shot-peened Case

Shot-peening is an operation that is used largely in the manufacture of mechanical parts. The technique consists of propelling at high speed small beads of steel, glass, or ceramic against the part to be treated. The size of the beads can vary from 0.1 to 1.3 or even 2 mm. The material used is a quenched and tempered 35NCD16 grade alloy steel. Chemical compositions and tensile properties are shown in Tables 1 and 2, respectively.

The main advantage of this surface treatment is that it considerably increases the fatigue life of mechanical parts by creating beneficial residual surface stresses. Unfortunately, the residual stress levels and distribution are generally altered when parts are subjected to fatigue loading. The problem is then to find out what the magnitude and distribution of the "stable" residual stresses are, and to include them in the design calculations. It is essential, therefore, to appreciate the stability of residual stresses as a function of the number of fatigue cycles, the cyclic stress amplitude, the ratio $R = \sigma_{min}/\sigma_{max}$, and the loading direction.

Application of the Calculation Software on the Shot-peened Case

For the calculation of the stabilized residual stress during fatigue, the software (PC2D) is used. We must introduce the initial plastic strains that correspond to the residual stresses. These plastic strains can be measured by the different techniques [*7,8*] or calculated by a prediction method [*9*]. In this paper, the residual stresses are measured by the X-ray diffraction method. The experimental results of the fatigue committee of the French Society for Metallurgy (SFM) are used [*10*].

The cyclic stress-strain curves are determined by constant strain amplitude loading during low-cycle fatigue tests. The evolution of the cyclic hardening properties in terms of the number of fatigue cycles is indicated in Fig. 5. The cyclic yield stress decreases with the number of cycles. Details of the peening parameters are shown in Table 3.

The residual stress distributions are measured before and after fatigue testing by the X-ray diffraction method for the different types of loading cases. The results are shown in Fig. 6.

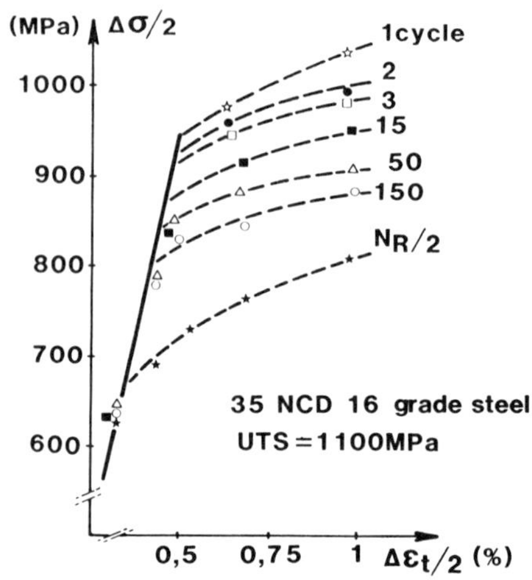

N (cycles)	Y.S. (MPa)
☆ 1	940
● 2	920
□ 3	910
■ 15	870
△ 50	830
○ 150	800
★ $N_R/2$	660

FIG. 5—*Evolution of the cyclic hardening properties in terms of number of cycles of 35NCD16 (U.T.S = 1100 MPa). The* ($N_r/2$) *curve corresponds to the stabilized stress measured for half-life of the specimens.* ($\Delta\sigma/2$ *and* $\Delta\epsilon/2$ *are the true stress amplitude and true plastic strain amplitude, respectively*).

The plastic initial strain distribution is calculated from the measured residual stresses. Suppose that there are no real strains (ϵ_o^r) on the specimen surface. It is equivalent to consider that the radius of the test specimen is semi-infinite in comparison with the depth of the stressed layer caused by the shot-peening treatment. For the initial state:

$$\epsilon_o^r = \epsilon_o^p + \epsilon_o^e = 0 \tag{6}$$

so we obtain:

$$\epsilon_o^p = -\epsilon_o^e$$

where

ϵ_o^e = elastic strains corresponding to the initial residual stresses, and
ϵ_o^p = initial plastic strains.

Using the elastic relations, we can obtain:

$$\epsilon_{zz}^p = -\frac{1}{E}\sigma_{zz} + \frac{\nu}{E}\sigma_{\theta\theta} \tag{7}$$

$$\epsilon_{\theta\theta}^p = -\frac{1}{E}\sigma_{\theta\theta} + \frac{\nu}{E}\sigma_{zz} \tag{8}$$

TABLE 3—*Shot peening parameters.*

Almen Intensity	Diameter of the Shot, mm	Coverage	Shot
F30-F35A (12-14A)	0.57	200%	S230

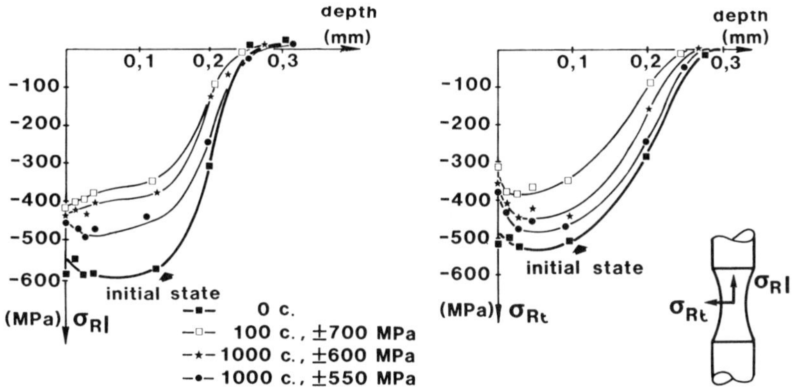

FIG. 6—*Evolution of axial* (σ_{Rl} *and tangential* (σ_{Rt}) *residual stress distribution during the tension-compression fatigue tests.*

$$\epsilon_r{}^p = -\frac{1}{E}(1 - \nu)(\sigma_{zz} + \sigma_{\theta\theta}) \tag{9}$$

where

$\sigma_{zz} = \sigma_{Rz}$ = axial residual stress, measured by X-ray diffraction,
$\sigma_{\theta\theta} = \sigma_{Rt}$ = tangential residual stress, measured by X-ray diffraction,
$\epsilon_{zr}{}^p = 0$ = absence of shear stress,
$\sigma_{rr} = 0$ = this hypothesis is valid at the surface, and
$\epsilon_{zz}{}^p + \epsilon_{\theta\theta}{}^p + \epsilon_{rr}{}^p = 0$ = plastic incompressibility of the material.

To verify these boundary conditions, the calculated plastic strains are introduced into the calculation software for a no-loading case. Very similar initial residual stress distributions are found in comparison with the initial measured residual stress. So we can suppose that the method of calculation of the initial plastic strains is valid.

Calculated Results and Experimental Results

Figures 7 and 8 show the calculated results of the state of the residual stress after one load cycle for three different stress amplitudes. We can see that the relaxation of residual

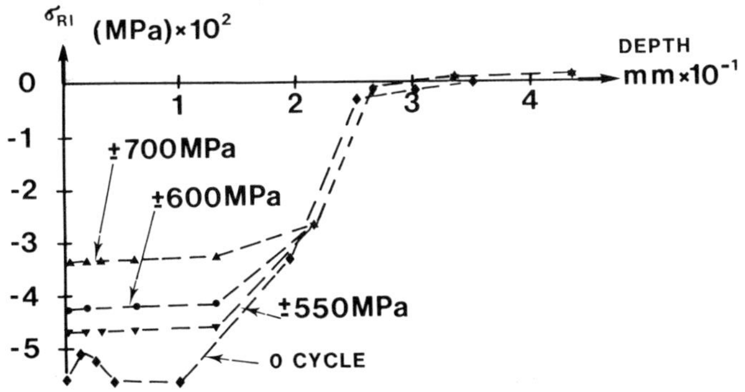

FIG. 7—*Calculated relaxation of axial residual stress after 1 cycle (influence of the stress amplitude).*

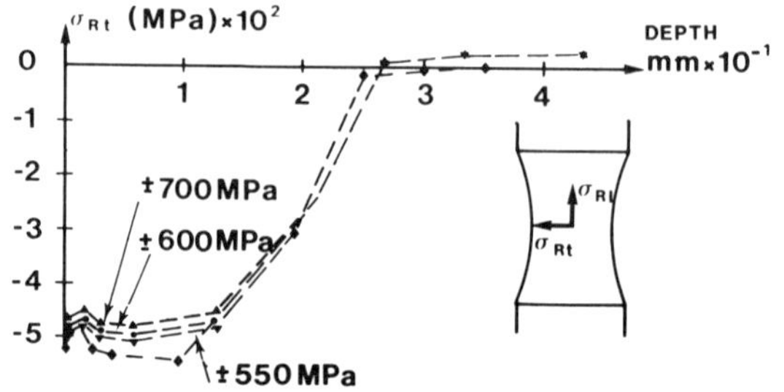

FIG. 8—*Calculated relaxation of tangential residual stress after 1 cycle (influence of the stress amplitude).*

stress increases with the level of the stress amplitude when there is a fully reversed fatigue loading. The direction of the applied load also influences the residual stress relaxation. In the loading (axial) direction (Fig. 7) the relaxation is greater than in the tangential direction (Fig. 8).

Figures 9 and 10 show the influence of number of the fatigue cycles on the calculated residual stress distribution. We can observe that the level of maximum compression decreases when the number of cycles increases. In this case, the material is cyclically softened. So the true yield stress decreases with the number of cycles. After each cycle, the material must reach a new stabilization situation caused by the decrease of the material cyclic properties. After $N_r/2$ cycles, the residual stress is stabilized.

Figures 11 and 12 show the effect of the level of the maximum compressive stress on relaxation for a fixed maximum stress (σ_{max} = 650 MPa). The stress relaxation increases with the level of the compressive stress of the fatigue loading. This behavior is natural because the initial residual stresses are compressions, so these stresses relax in the compression phase. For the case of shot-peening, the compressive level of the fatigue loading is a very important factor. For σ_m (mean stress) = 0, a very large relaxation is observed.

An examination of Fig. 13 shows that for a constant stress amplitude, the relaxation of the shot-peened residual stress decreases with the increase of the mean stress. Here the parameter R ($\sigma_{min}/\sigma_{maxi}$) is studied. For the case of a high mean stress (σ_m = 150 MPa) σ_{Rl} and σ_{Rt} are similar. (See Fig. 14.)

Figures 15, 16, and 17 show the comparison between the results measured by the X-ray diffraction method and the results calculated by the finite element method. These results

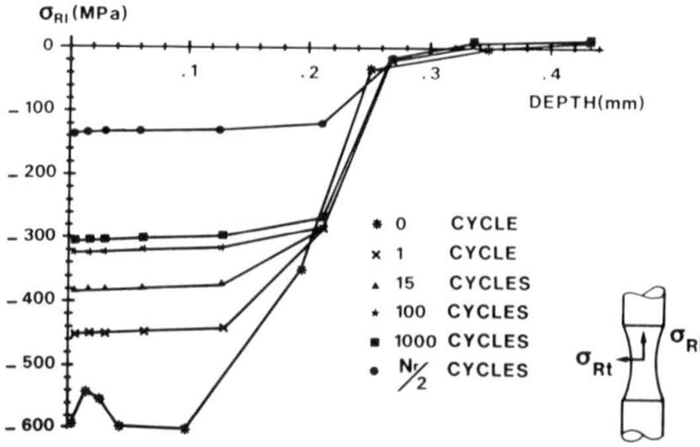

FIG. 9—*Calculated evolution of axial residual stress distribution with the number of cycles for a constant applied stress of ±600 MPa.*

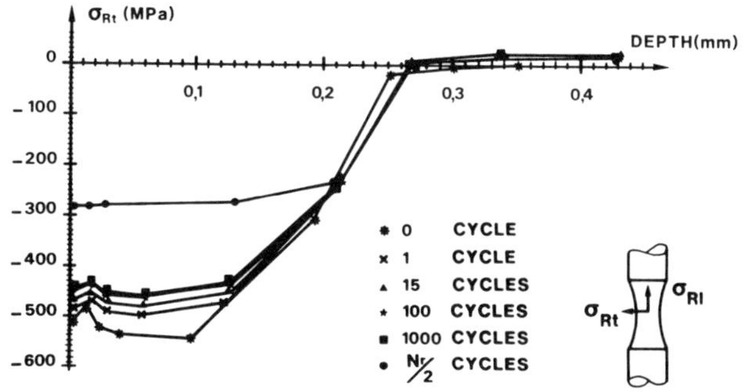

FIG. 10—*Calculated evolution of tangential residual stress distribution with the number of cycles for a constant applied stress of ±600 MPa.*

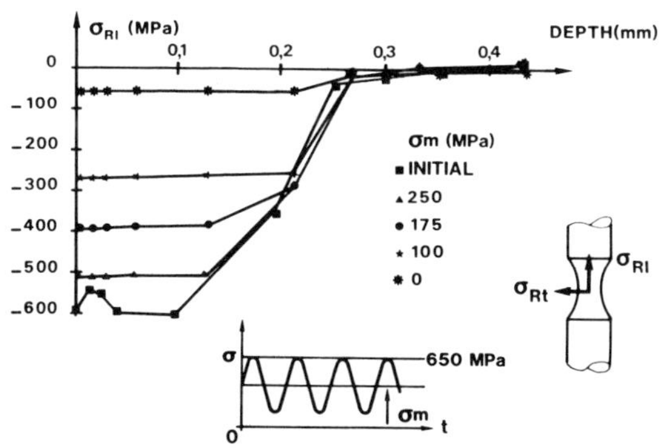

FIG. 11—*Calculated evolution of axial residual stress relaxation when the level of the applied compression changes for a case of a fixed applied maximum stress* (σ_{max} = *650 MPa) after* N$_r$/2 *fatigue cycles.* σ_m *is the mean stress of the fatigue loading.*

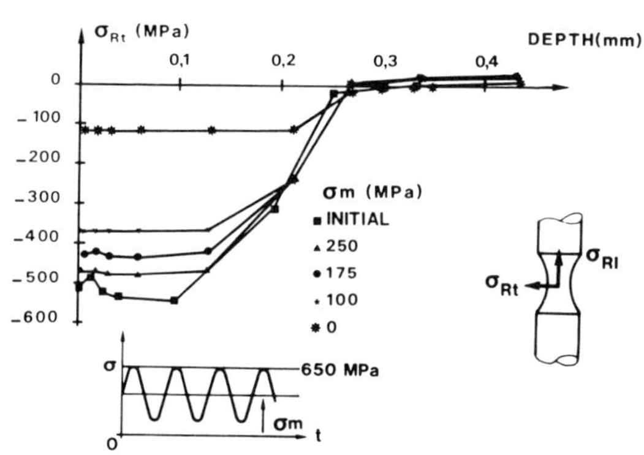

FIG. 12—*Calculated evolution of tangential residual stress relaxation with the same loading conditions as Fig. 11.* σ_m *is the mean stress of the loading fatigue.*

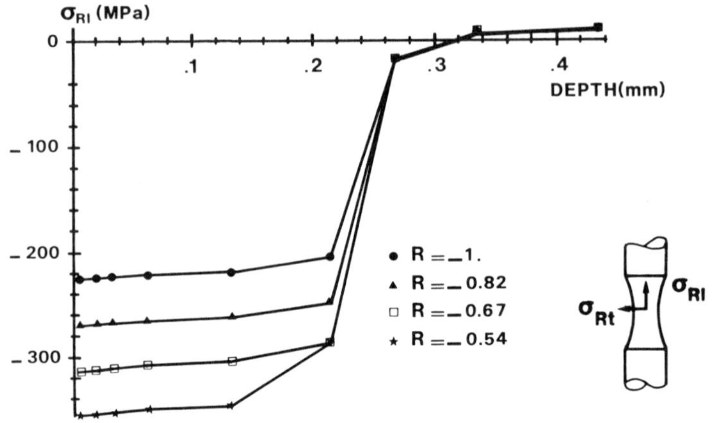

FIG. 13—*Calculated effect of ratio, R,* ($\sigma_{min}/\sigma_{max}$) *on relaxation of the axial residual stress with a constant stress amplitude* ($\Delta\sigma$ = *1000 MPa*) *fatigue after* $N_r/2$ *fatigue cycles.*

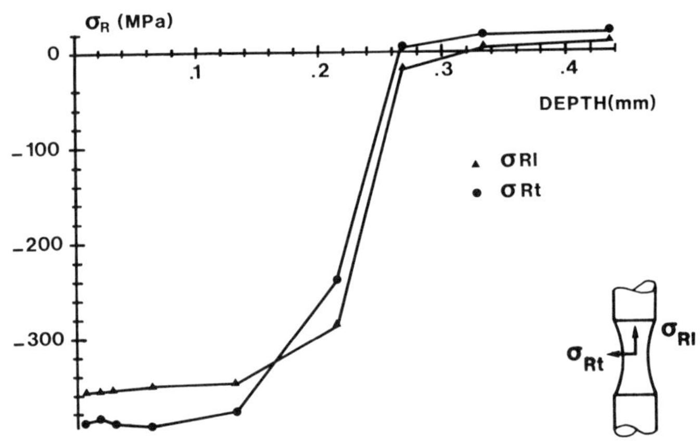

FIG. 14—*Stabilized residual stress after tension-compression fatigue testing* (R = -0.54, σ_{max} = *650 MPa*).

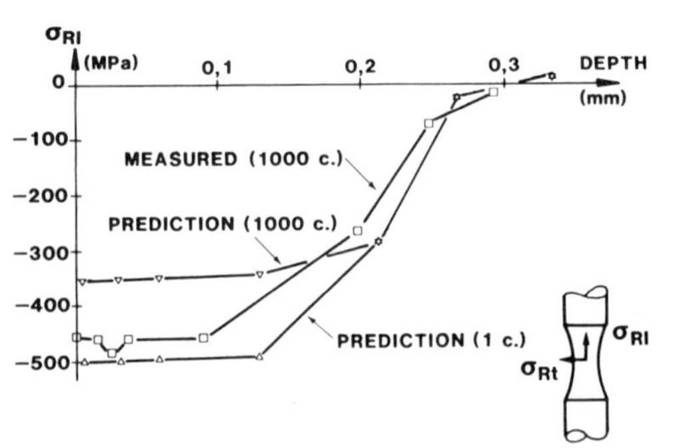

FIG. 15—*Comparison of the calculated and measured results of the relaxation of axial residual stress* ($\pm$*550 MPa*).

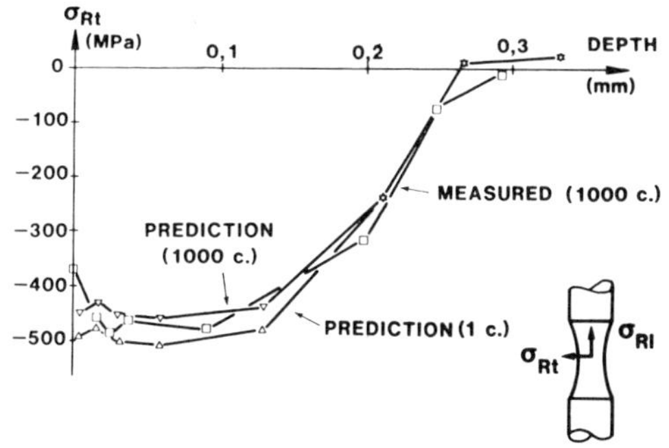

FIG. 16—*Comparison of the calculated and measured results of the relaxation of tangential residual stress* (±*550 MPa*).

show that the finite element method predicts the same phenomenon as the experimental observations for the effects of the stress amplitude. For the axial stress, the calculation model overestimates the relaxation of the residual stress using the cyclical properties of the material. For tangential stress, the model and the experimental results give the same results. If we compare the calculated results using the mechanical properties at 1 cycle and at 1000 cycles with measured results after 1000 cycles of fatigue loading (Figs. 15 and 17), we can observe that the measured stresses are between the two calculated results. The difference between the model and the measurement can be explained by the following causes:

1. The cyclical properties of material without the shot-peening are used in the calculation. But there are two layers of material, the shot-peened layer and the non shot-peened layer. A difference of the static mechanical properties between two layers was observed [*11*]. Probably the cyclic properties also may be different.
2. At 1000 cycles, the material is not in a stabilized state. For this state, the tensile and the compression properties can be different.
3. The initial residual stress can have some dispersions for the different specimens used in the experiment because the method of the measurement of the residual stress distribution by the X-ray method is a destructive method due to electrolytical machining of the specimen. The measurements of the residual stress before and after fatigue testing were made on different specimens.

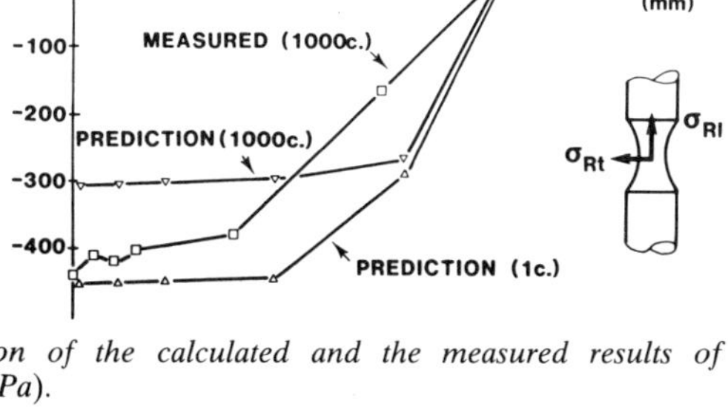

FIG. 17—*Comparison of the calculated and the measured results of the relaxation of axial residual stress* (±*600 MPa*).

TABLE 4—*Chemical composition of the 42CD4 grade steel.*

C	Si	Mn	S	P	Ni	Cr	Mo
0.43	0.30	1.02	0.021	0.015	0.19	0.94	0.20

Ground Case

Grinding is one of the most used modes of finishing in mechanical engineering. It is very important to study the residual stresses introduced by this machining method, because it often determines the final state of residual stresses in a part, which influences its behavior in fatigue service.

A detailed study of the residual stress due to grinding is outside the scope of this paper. We shall just analyze the relaxation of residual stress during fatigue testing. The simplified finite element method will be used for the prediction of this phenomenon and the calculation results will be compared with the results of measurement using the X-ray diffraction method. The material used is a quenched and tempered 42CD4 grade alloy steel. Chemical compositions, tensile properties, and the heat treatment are shown in Tables 4, 5, and 6, respectively.

The cyclic stress-strain curves are determined by a conventional method (one specimen for one strain level) (Fig. 18). We can see that it is a cyclically softening material.

In the calculation model, we studied the reversed plane bending case. The assumption of a plane strain state is used:

$$\epsilon_{zz} = 0$$

Figure 19 shows the loading mode used in fatigue testing.

The initial residual stresses are measured by the X-ray diffraction method. The initial plastic strain, $\epsilon_{ij}{}^{p}$, can be calculated by the same method as the shot-peened case. We can obtain the following:

$$\epsilon_{xx}{}^{p} = \frac{1}{E}(-\sigma_{xx} + \nu\sigma_{zz}) \tag{10}$$

$$\epsilon_{yy}{}^{p} = \frac{1-\nu}{E}(\sigma_{xx} + \sigma_{zz}) \tag{11}$$

$$\epsilon_{zz}{}^{p} = \frac{1}{E}(\nu\sigma_{xx} - \sigma_{zz}) \tag{12}$$

where

$\sigma_{xx} = \sigma_{Rl}$ = longitudinal residual stress,
$\sigma_{zz} = \sigma_{Rt}$ = transverse residual stress,
σ_{yy} = 0, and
σ_{xy} = 0.

TABLE 5—*Tensile properties of 42CD4 grade steel.*

Yield Stress, MPa	Ultimate Tensile Strength, MPa	Elongation, %	Reduction in Area, %
1180	1250	7	33

TABLE 6—*Heat treatment and hardness of 42CD4 grade steel.*

Heat Treatment	Hardness, HRC
Quench after heating at 850°C, 30 min, tempering at 520°C, 1 h	40

The program tested for the relaxation by fatigue of residual grinding stresses gave satisfactory results. Figure 20 shows the results obtained for hard longitudinal grinding. The predicted reduction agrees fairly well with the experimental results in the case of hard grinding, for which the distribution curve of the residual stresses can be measured with fair accuracy. Agreement between the model and the experimental results is worse for soft grinding (Fig. 21) for which the determination of the stress gradient before and after fatigue is more imprecise. On the other hand, the model of the cyclic behavior of the layer very close to the surface cannot be represented by the macroscopic cyclic work-hardening curve of a test specimen because the microstructures are often different. This might be the modeling limit for a microscopic localized phenomenon using the rules of macroscopic behavior and calculation.

Conclusion

In the work completed, a computer program of the relaxation model using the finite element method to calculate the stabilized residual stress during cyclic loading is used. For a small radial cyclic loading, elastic calculations for only a few cycles are necessary to find the stabilized state (elastic shakedown).

This method is applied to shot-peened and ground cases. The different fatigue parameters often used in material research are studied, such as the number of cycles, the stress amplitude, and the ratio R ($\sigma_{min}/\sigma_{max}$). The result is satisfactory enough for the study of the residual stress relaxation phenomenon. It is a simple method, for the prediction of the stabilized multiaxial residual stress state in the depth plane. Only rarely is residual stress relaxation analyzed systematically due to the long experimental time for fatigue testing and the residual stress distribution measurement. This study shows a new way for understanding the mechanical relaxation of residual stress. However, for better modeling with this method it is

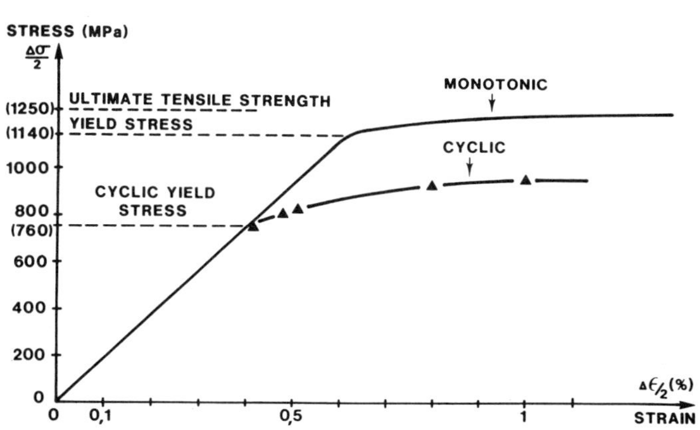

FIG. 18—*Monotonic and cyclic stress-strain curves for 42CD4 grade steel.*

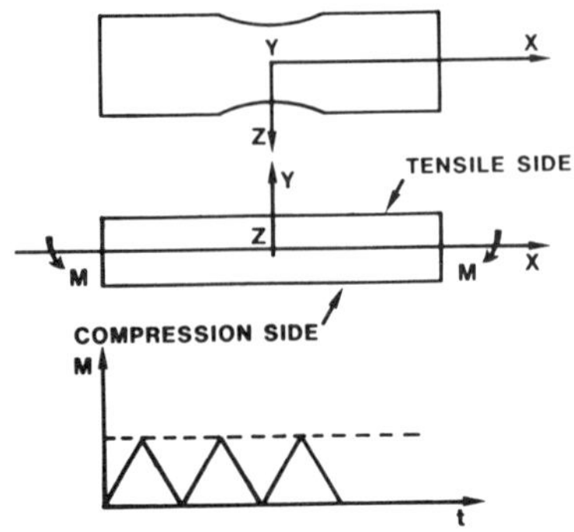

FIG. 19—*Loading mode of the specimen.*

necessary to introduce into the calculation the real cyclic stress-strain curve corresponding to the materials present in the prestressed layer. This represents a difficulty in the case, for example, of grinding or shot peening. Nevertheless, the real behavior of the prestressed layer could be identified and adjusted by comparison of calculated and measured results obtained for one loading. Then the model can be applied for other loading cases.

In our approach, only the macrostress effect is considered. In fact, the relaxation of residual stress is the consequence of the dislocation arrangements that depend upon the macroplastic and microplastic strain. The mobility of the dislocations depend greatly upon the initial dislocation density and the arrangement which depend upon the origins of the material (cold worked, machined, or heat treated). The relaxation by macroplastic strain is a dominant phenomenon for a high-applied stress amplitude.

The cyclic properties of the material are also an important parameter. Using the mechanical approach, we can make the following general predictions:

1. For a cyclic hardening material, a large part of the relaxation of residual stress is realized in the first series of fatigue cycles. Because, for this case, the real yield limit increases with the number of loading cycles, so the radius of the limit surface corre-

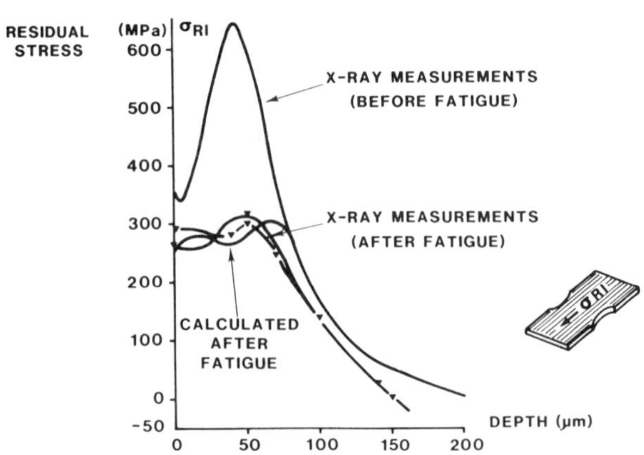

FIG. 20—*Predicted reduction of the residual stresses produced by hard longitudinal grinding. X-ray diffraction measurements made on two test specimens after fatigue testing.*

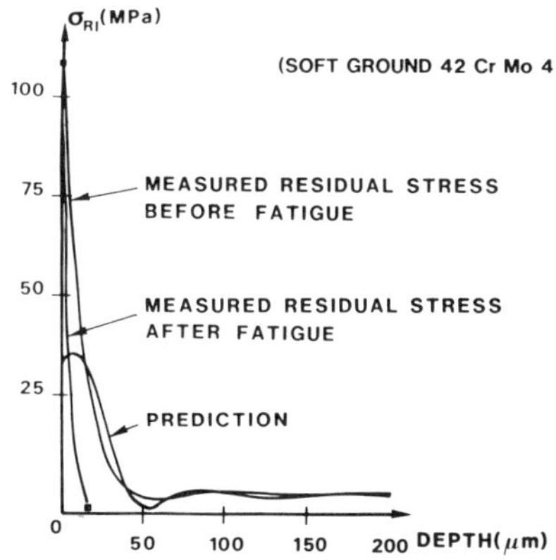

FIG. 21—*Comparison of the calculated and the measured results of the relaxation of axial residual stress* (σ_{RI}) *produced by soft longitudinal grinding case.*

sponding to the nature of the applied loading increases also. So we find the case of elastic shakedown after the first cycles.

2. For a cyclic softening material, the residual stresses relax with the increase of the number of cycles up to a stabilized state of cyclic properties.

Some experimental results confirm such predictions. The various work [*12*], [*13*] on the relaxation of residual stress on different aluminum alloys (AL7075-T7531 and AL2024-T351 [*12*], Al2129-T851 [*13*]) that are cyclic hardening materials show that the residual stress relaxes only in the first cycles. In this paper, we observed that the stabilization of the residual stress is very slow for two softening materials. These observations do not depend on the origins of the residual stress, for example, cold worked (shot-peening) and machined (milling [*12*] and grinding).

Nevertheless, we must be careful of the limitations of a simplified mechanical approach. Sometimes the residual stress can be relaxed when the local macrostress does not exceed the cyclic yield points for example, relaxation of residual stress by vibration. The mechanical approach must be limited to the cases in which the applied stresses are high (such as fatigue testing). The relaxation of residual stress is a very complex phenomenon that is influenced by many factors, such as the applied stress, plastic deformation, the number of cycles, the origins of residual stresses, and the material. With the calculated stabilized residual stress, the fatigue behavior of the material can be evaluated using multiaxial fatigue failure criteria, Dang Van, or Crossland criteria [*2*]. The real fatigue limit can also be estimated by these methods. In the future, experimental work will continue to examine the validity of this prediction method for other materials and treatments. Research on the quantitative relations between the movement of dislocations during the relaxation of residual stress and the cyclic properties of material, the applied stress and the nature of the material will lead to a better understanding of the residual stress relaxation phenomenon during fatigue.

References

[*1*] *Residual Stress Effects in Fatigue, ASTM STP 776,* 1982.

[*2*] Skalli, N. and Flavenot, J. F., "Prise en Compte des Contraintes Résiduelles dans un Calcul Prévisionnel de Tenue en Fatigue," *CETIM Informations,* No. 90, May 1985, pp. 35–47.

[*3*] Flavenot, J. F. and Bramat, M., "Influence de Differents Traitements de Parachè-vement sur la

Tenue en Fatigue de Pièces Soudées," *Proceedings of the Second International Conference on Shot Peening* (*ICSP-2*), H. O. Fuchs, Ed., Chicago, USA, 1984, pp. 355–360.

[*4*] Picouet, L., Bigonnet, A., Lieurade, H. P., and Castex, L., "Influence du Grenaillage sur le Comportement en Fatigue d'Assemblages Soudés en Acier HLE," *Proceedings of the Second International Conference on Shot Peening* (*ICSP-2*), pp. 346–354.

[*5*] Zarka, J. and Casier, J., "Elastic-plastic Response of a Structure to Cyclic Loadings: Practical Rules," *Mechanics Today,* Vol. 6, 1981, pp. 93–198.

[*6*] Lemaitre, J. and Chaboche, J. L., *"Mécanique des Matériaux Solides,"* DUNOD, France, 1985.

[*7*] Lu, J., Niku-Lari, A., and Flavenot, J. F., "Mesure de la Distribution des Contraintes Résiduelles en Profondeur par la Méthode du Trou Incrémentale," *Mémoires et Études Scientifiques de la Revue de Métallurgie,* February 1985, pp. 69–81.

[*8*] Maeder, G., "Mesure des Contraintes Résiduelles par Diffraction X, Applications," *Revue Française de Mécanique,* No. 82, 1982, pp. 57–70.

[*9*] Guechichi, H., Castex, L., Frelat, J., and Inglebert, G., *"Predicting Residual Stresses due to Shot Peening," Proceedings of Tenth Conference on Shot Peening* (*CETIM-ITI*), Senlis, France, Nov. 1986, pp. 23–34.

[*10*] Bignonnet, A., "Tenue à la Fatigue d'un Acier 35NCD16 Grenaillé, Évolution des Contraintes Résiduelles de Grenaillage en Fonction du Type de Sollicitation," *Proceedings of Ninth Conference on Shot Peening* (*CETIM*), Senlis, France, Nov. 1985, pp. 107–116.

[*11*] Picouet, L., "Fatigue des Assemblages Soudés," doctoral thesis, ENSAM, Paris, France, September 1985.

[*12*] Bonnafe, J. P., "Etude par Diffractométrie X et Émission Acoustique de l'Endommagement par Fatigue des Alliages d'Aluminium à haute Résistance Grenaillés," doctoral thesis, ENSAM, Paris, France, May 1986.

[*13*] James, M. R. and Morris, W. L., "The Relaxation of Machining Stresses in Aluminium Alloys during Fatigue," *Residual Stress for Designers and Metallurgists*, ASM, 1981, pp. 169–188.

J. F. Flavenot and N. Skalli[1]

Effects of Grinding Conditions on Fatigue Behavior of 42CD4 Grade Steel; Comparison of Different Fatigue Criteria Incorporating Residual Stresses

REFERENCE: Flavenot, J. F. and Skalli, N., "**Effects of Grinding Conditions on Fatigue Behavior of 42CD4 Grade Steel; Comparison of Different Fatigue Criteria Incorporating Residual Stresses,**" *Mechanical Relaxation of Residual Stresses, ASTM STP 993,* L. Mordfin, Ed., American Society for Testing and Materials, Philadelphia, 1988, pp. 91–111.

ABSTRACT: Fatigue tests carried out at Centre Technique des Industries Mécaniques (CETIM) on a NF 42CD4 grade steel (G 41420) show that grinding parameters such as down feed and the type of wheel have a very clear influence on residual stresses and fatigue limit under plane bending. The stability in fatigue of residual stresses obtained by three grinding conditions were evaluated, using X-ray diffraction methods, in order to introduce values of the stabilized residual stresses into a calculation procedure developed for determining fatigue strength. This analysis contributes to a better knowledge of grinding effects in fatigue and allows the distinction between the respective influence of surface roughness and residual stresses. This project also demonstrates the importance of the reduction of the tensile residual stresses during fatigue, and the relationship between this reduction and the behavior of the metal under cyclic loading.

KEY WORDS: residual stress, relaxation, grinding, fatigue, fatigue criterion

Introduction

Grinding is a machining operation that can be used for the finishing of many parts which require close tolerances, or when the mechanical properties of the steel being used are high. If a poor choice of grinding wheel is made, or if the grinding conditions are too severe, the operation generally results in the generation of heat sufficient enough to generate residual stresses in the surface layers of the metal. The behavior of ground parts in service is dependent not only on the surface roughness, but also on the residual stresses.

From the results of work stated in the references and from fatigue testing carried out at CETIM [*1*] on parts that have been turned or ground, a nomogram was produced from which the coefficient of reduction in fatigue limit, as a function of the surface roughness and ultimate strength of the steel, can be obtained (Fig. 1). Unfortunately, this type of nomogram, of which there are many to be found in the references [*1*], takes into account both the influence of the surface roughness and the influence of the residual stresses on the reduction in fatigue strength. In this type of nomogram the respective effects of the roughness and residual stresses are mixed.

In addition, other tests carried out on 42CD4 steel show that the residual stresses due to grinding can change considerably as the grinding conditions change (type of grinding wheel,

[1] Engineers, Centre Technique des Industries Mécaniques (CETIM), Senlis, France 60300.

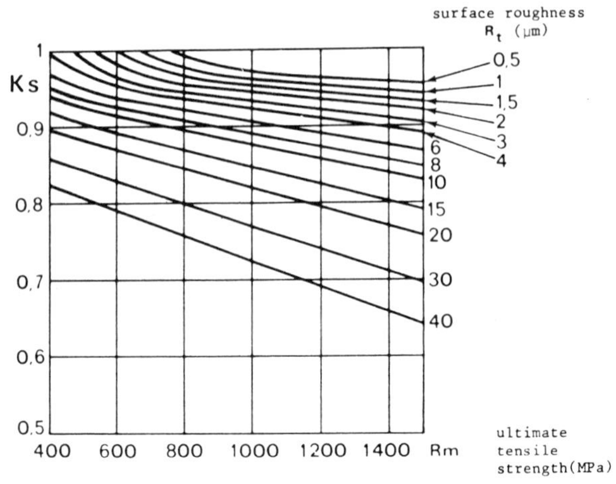

FIG. 1—*Chart of K_s versus surface roughness and ultimate tensil strength.*

down-feed). These tests show that it is not possible to correlate the surface roughness with the induced residual stresses [*2*].

In order to estimate the influence of a machining method on the fatigue strength, it appears to be necessary to separate the respective effects of roughness and residual stresses. For this reason, in this investigation, greater attention is paid to the residual stresses: to their change with the machining parameters, to their reduction during fatigue tests; and to their incorporation in a method of calculation of the fatigue strength. From fatigue tests in repeated plane bending (constant amplitude loading with stress ratio $R = 0.05$) of 42CD4 steel, it is possible to show that the initial residual stresses are partially reduced during fatigue testing. It is also possible to introduce these residual stresses into fatigue criteria in order to predict the fatigue strength of the ground part. Lastly, the results obtained allow the respective influences of residual stresses and surface roughness on the fatigue limit to be separated from each other.

Test Program, Steel, and Specimen Manufacture

The 42CD4 steel chosen for this investigation is widely used in mechanical engineering. Its high mechanical properties and its operating conditions often require finishing by grinding.

Tables 1 and 2 give the chemical composition and the mechanical properties after quench and tempering treatment (30 min austenitizing at 850°C (1587°F) and 30 min tempering at 520°C (993°F)).

The test specimens were machined from flat rolled section. First their thickness was reduced to 6 mm by milling and then they were subjected to the heat treatment described above. After heat treatment, all the test specimens were soft ground to a thickness of 5.5 + 0.01 mm. Then they were machined to the final dimension of 5 + 0.02 mm, in accordance

TABLE 1—*Chemical composition of the 42CD4 grade steel.*

C	Si	Mn	S	P	Ni	Cr	Mo
0.43	0.30	1.02	0.021	0.015	0.19	0.94	0.20

TABLE 2—*Mechanical properties of the 42CD4 grade steel.*

Yield Stress 0.02 %, MPa	Ultimate Tensile Strength, MPa	Elongation A, %	Reduction in Area, %	Hardness, HRC
1180	1250	7	33	40

with the procedures described in Table 3. The final geometry of the test specimens is shown in Fig. 2. For each case (fixed grinding wheel and finishing conditions), twenty test specimens were ground under the same conditions.

For a given grinding wheel, it was possible to adjust the down-feed only. In our tests, movements of the grinding wheel and the part were opposed. The other parameters, especially the speed of the grinding wheel and the speed of the part, were fixed on the grinding machine used.

- Speed of the grinding wheel: 30 m/s.
- Speed of the table: 13 m/min.
- Transverse speed of the grinding wheel (continuous): 7.7 mm/s.
- Length between edges: 200 mm.
- Lubricant: emulsifiable cutting oil.

Dressing conditions for the grinding wheel were as follows:

- The grinding wheel was dressed before each face of the test specimen was machined.
- Dressing was carried out in both directions.
- Down feed: 0.03 mm.
- Dressing speed: for 2850 rpm, Vd (peripheric speed) = 285 mm/min; fd (feed) = 0.1 mm/rev.

Figure 3 shows the microstructures resulting from the three grinding conditions studied. The surface roughness of the part subjected to hard grinding is altered. The scanning microscope provides a good idea of the different surfaces obtained from the three grinding conditions studied (Fig. 4).

TABLE 3—*Machining procedures used.*

Grinding Type	Mark	Wheel Type Young's Modulus, MPa	Finishing Process Depth of Cut	Roughness, R_t, μm
Soft longitudinal	AL	soft A46HV $E = 41.4 \times 10^3$	5	7.4
Soft transverse	AT	soft A46HV $E = 41.4 \times 10^3$	5	7.5
Medium longitudinal	CL	medium A48KV $E = 41.4 \times 10^3$	15	8.7
Severe longitudinal	DL	hard A3605V $E = 57.8 \times 10^3$	30	11.9
Severe transverse	DT	hard A3605V $E = 57.8 \times 10^3$	30	11.4

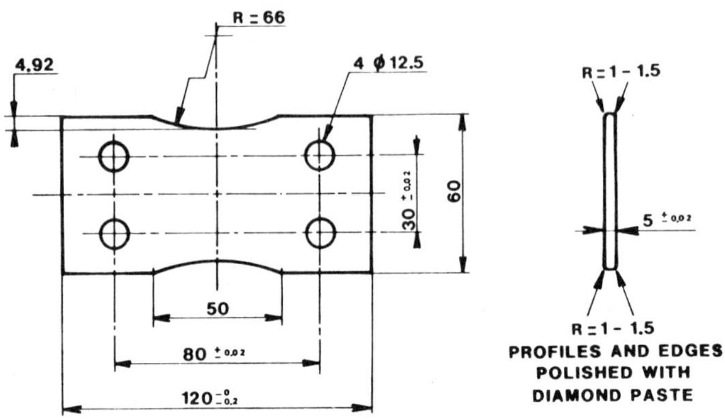

FIG. 2—*Geometry of test specimen for repeated plane bending tests, dimensions in millimeters.*

Fatigue Testing

The fatigue tests were carried out on an electrohydraulic machine, usually used for torsion fatigue tests. A special fixture allowed alternating or repeated bending tests to be performed on the flat test specimens (Fig. 5).

The fatigue strength in repeated plane bending at 5 million cycles was determined by the standard staircase method [*3*]. The results obtained in terms of maximum stresses, are given in Table 4:

A very distinct reduction in fatigue life (−26%) is noted for hard grinding, when compared to soft grinding. The soft and medium machining conditions give a similar fatigue limit. The orientation of the grinding marks in relation to the axis of the applied stress also affects fatigue life. As might be expected, transverse grinding always gives a lower fatigue life than longitudinal grinding. For the soft conditions, the fatigue strength for the transverse grinding is statistically the same number as for the longitudinal grinding.

Fracture examination with a scanning electron microscope did not allow accurate determination of the depths at which fatigue cracks initiated. They probably initiate at the surface, at a depth of 10 to 20 μm (Fig. 6).

Measurement of Residual Stresses—Variation of the Stresses with Machining Conditions

Residual stresses were measured on the ground specimens using the standard X-ray diffraction method. When residual stresses due to grinding are measured by X-ray diffraction along the grinding marks, the presence of a shear stress can be seen in a direction parallel to the grinding marks. This results in an ellipse when one attempts to relate the angle of diffraction of the X-rays with the function $\sin^2\psi$ (Fig. 7). Details of this method are described in [*4*].

The presence of the shear stress in the direction parallel to the grinding marks requires twice the number of measurements (ψ positive and ψ negative) of the perpendicular case in order to correctly determine the equation of the ellipse from which the residual stress is calculated.

The measurements carried out at two points (one at the center and one at the edge of specimen) on three hard-ground test specimens reveal a scatter in the values of the residual stresses, particularly in the first few microns of the surface layer. After the peak stress, this scatter becomes negligible compared to the scatter of accuracy of the measurement itself.

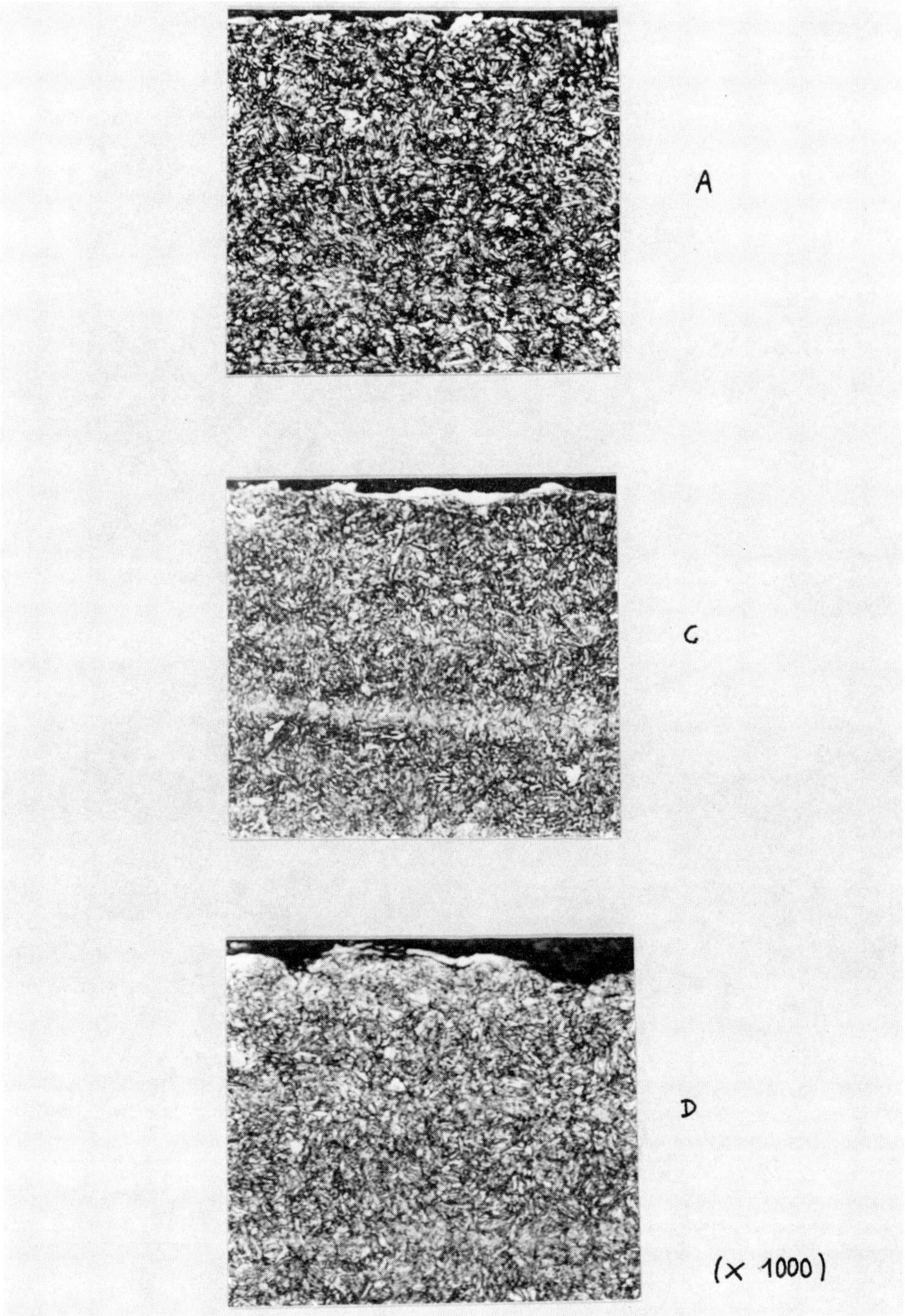

FIG. 3—*Microstructure of the ground surface layers for soft (A), medium (C), and severe (D) grinding.*

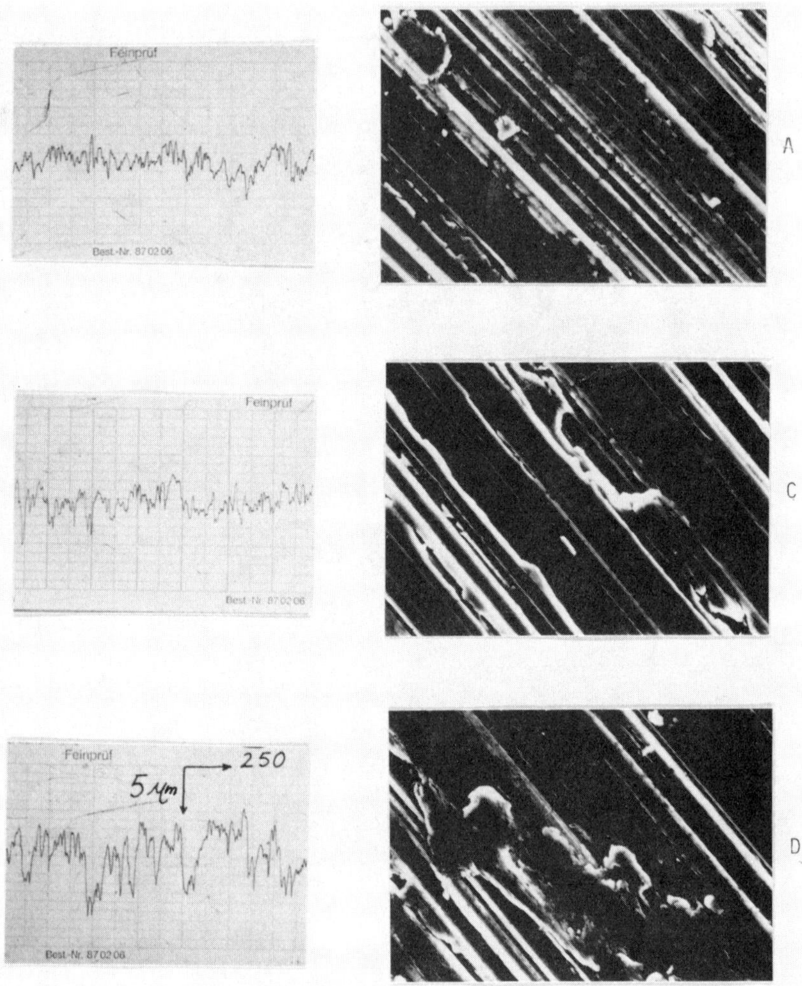

FIG. 4—*View of the surface after grinding using A, C, or D conditions (scanning electron microscope examination and surface roughness profiles in micrometres).*

TABLE 4—*Fatigue test results.*

Type of Grinding	Fatigue Strength and Standard Deviation, MPa	Roughness, R_t, μm
Soft longitudinal	762 ± 102	7.4
Soft transverse	690 ± 105	7.5
Medium longitudinal	751 ± 8	8.7
Severe longitudinal	564 ± 13	11.9
Severe transverse	520 ± 41	11.4

FIG. 5—*View of fatigue test fixture.*

Despite the scatter observed, a mean distribution curve of residual stresses may be plotted very easily (Fig. 8).

The measurements of the profile of the residual stress versus the depth need step-by-step material removal (electrochemical removal). Two residual stresses are measured: the longitudinal residual stress, σR_l, parallel to the grinding marks; and the transverse residual stress, σR_t, perpendicular to the grinding marks.

Figure 9 shows the distribution curves for longitudinal (σR_l) and transverse (σR_t) residual stresses before fatigue for a test specimen that was hard-ground. It is noted that the peak transverse residual stress is some 10 to 15 % lower than that of the longitudinal residual stress, and that their locations are at the same depth. During this residual stress measurement it is observed that the depth of the tensile residual stress peak corresponds to that where the shear residual stress referred to in Fig. 7 becomes effectively zero.

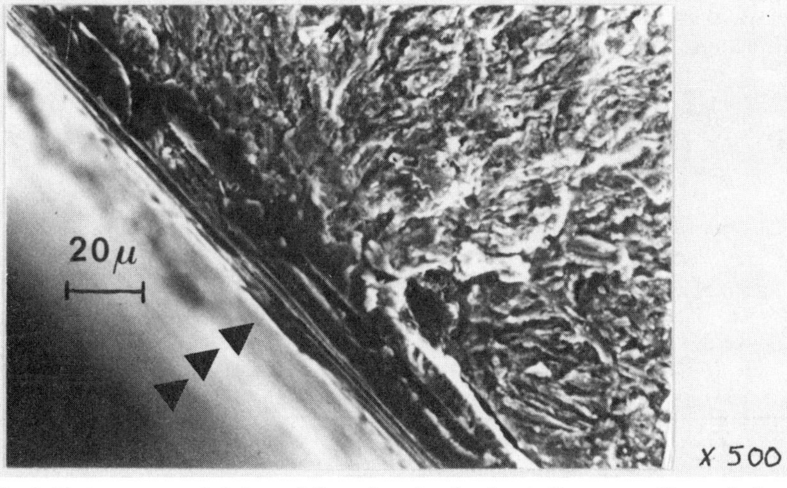

FIG. 6—*Initiation zone of fatigue failure for the hard grinding case. View of the fractured surface.*

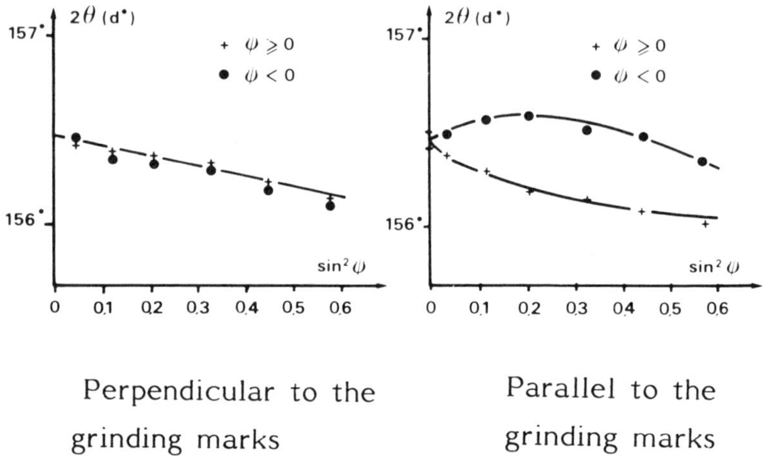

FIG. 7—*Effect of the direction of measurements on the relationship* $2\theta = f(\sin^2 \psi)$. (ψ *represents the angle between the normal to the crystallographic planes observed and the normal to the part and* θ *is the diffraction angle.*)

For the different grinding conditions chosen, distinctly different distribution curves of the residual stresses were obtained (Fig. 10). Hard grinding, either longitudinal (condition DL) or transverse (reference DT), resulted in particularly high residual stresses. No significant difference was observed between the two directions of grinding, longitudinal or transverse. In both cases, the peak residual stress transverse to grinding direction is lower by 10 to 15 % to the peak residual longitudinal stresses. For soft (condition A) and medium (condition C) grinding conditions, only very low levels of residual stresses were observed.

Relaxation of Residual Stresses After Fatigue Testing

The relaxation of residual stresses by fatigue testing was checked by measuring them with X-ray diffraction. Residual stresses were measured on parts that did not fail in fatigue, and

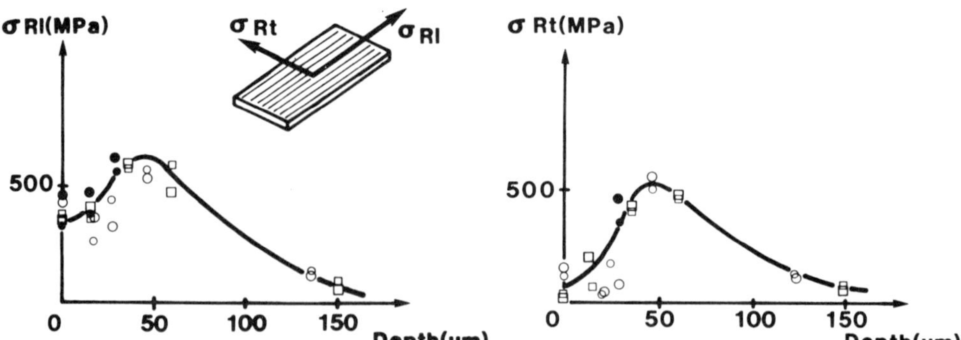

FIG. 8—*Scatter in the values of the residual stresses in the case of severe grinding. (D conditions: ● first, ○ second, □ third sample.)*

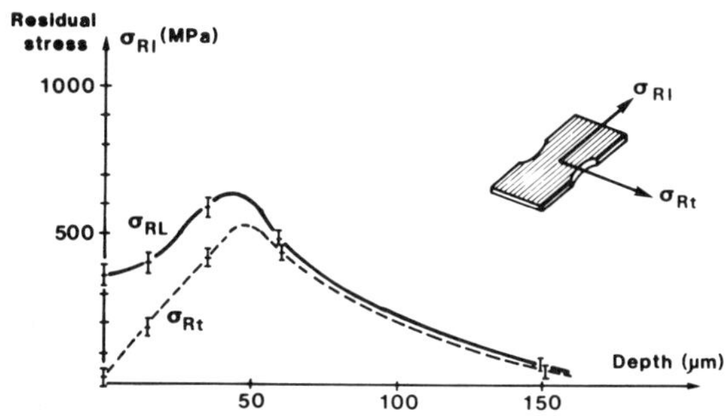

FIG. 9—*Curves of longitudinal and transverse residual stresses for hard-grinding.*

on those that tested close to the fatigue limit. The reduction in residual stresses observed corresponds to a loading close to the fatigue limit of the part.

The measurements were made on the face of the test specimen plane subjected to tensile stressing during repeated bending. The material was subjected to a minimum fatigue stress equal to the residual stress (σR) and to a maximum fatigue stress of $\sigma f_{max} + \sigma R$. The values found before and after the fatigue test are given in the Table 5.

The following deductions may be made from the results in Table 5.

For hard longitudinal grinding (condition DL):

- The longitudinal residual stress in the direction of applied load is reduced by 52% while the transverse residual stress is reduced by only 43%.
- The prestressed layer depth does not vary (Fig. 11).
- The stress relaxation occurs principally in the area between the surface and the peak stress (Fig. 11). It is in this area that the maximum stress is reached during fatigue cycling. Plastic micro-strains probably occur here, due to the elastic limit of the material being exceeded locally, and are the origin of the stress relaxation.
- The distribution of the stresses after fatigue tends to be more regular. For the longitudinal stress, the stress relaxation is higher. The peak stress tends to diminish to the value of the residual stress at the surface, resulting in a reduction in the stress gradient.

For hard transverse grinding (condition DT):

- In this case, it is the transverse stress (σR_t) that is reduced the most, since it is parallel to the applied load (Fig. 12).

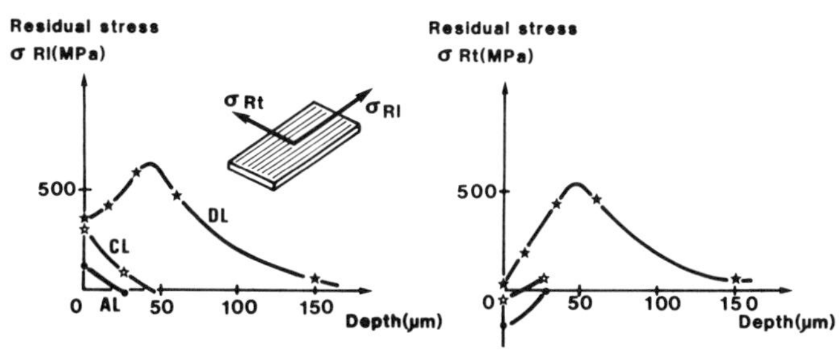

FIG. 10—*Distribution curves of residual stresses for different grinding conditions.*

TABLE 5—*Fatigue strength and residual stresses obtained before fatigue (for $\sigma f_{max} = 0$) and after fatigue testing at σf_{max}.*

Machining Conditions	Test Specimen Number	σf_{max}, MPa	Longitudinal Residual Stress: $\sigma R_{l\,max}$, MPa	Longitudinal Residual Stress: $\sigma R_{l\,max}$, MPa, mean value	Transverse Residual Stress: $\sigma R_{t\,max}$, MPa	Transverse Residual Stress: $\sigma R_{t\,max}$, MPa, mean value	Relaxation, %: $\frac{\Delta\sigma R_l}{\sigma R_l}$	Relaxation, %: $\frac{\Delta\sigma R_t}{\sigma R_t}$
AL, longitudinal soft grinding	A5	0	118	...	−178	...	...	...
	A13	762	113	...	−157	...	...	...
AT, transversal soft grinding	A16	690	125	...	−200	...	...	...
CL, longitudinal medium grinding	C5	0	307	...	−31	...	23	...
	C22	751	233	237	−122	...	23	...
	D20	751	241	237	−13	...	23	...
DL, longitudinal severe grinding	D10	0	640	630	603	566	52	43
	D10	0	630	630	570	566	52	43
	D20	0	620	630	540	566	52	43
	D20	0	630	630	550	566	52	43
	D47	564	337	300	376	320	52	43
	D8	564	250	300	272	320	52	43
	D8	564	309	300	311	320	52	43
DT, transversal severe grinding	D73	0	600	610	550	550	43	48
	D73	0	620	610	550	550	43	48
	D58	520	361	350	260	284	43	48
	D92	520	350	350	285	284	43	48
	D92	520	340	350	308	284	43	48

σf_{max} = maximum fatigue stress applied during fatigue testing (5.10^6 cycles),
$\sigma R_{l\,max}$ = peak value of the longitudinal residual stress, and
$\sigma R_{t\,max}$ = peak value of the transverse residual stresses.

NOTE—Relaxation, percent, is calculated using residual stresses obtained before fatigue (for $\sigma f_{max} = 0$ in the table) and after fatigue.

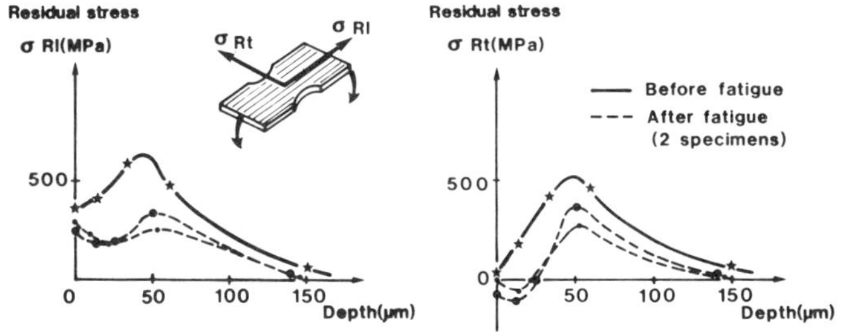

FIG. 11—*Residual stresses before and after fatigue testing in the case of hard grinding (condition DL). For the measurements after fatigue testing, the curves shown were obtained from two specimens.*

For soft and medium grinding:

- After fatigue testing, a reduction of the initial stresses was observed.
- Despite the low stress levels measured before and after fatigue testing, a change of stresses may be evidenced (Figs. 13 and 14).
- Due to the scatter in the stress measurements in the layers very close to the surface, these results must be considered only as reasonable orders of magnitude.
- It is noted that for soft and medium grinding, a tensile longitudinal residual stress and a compressive transverse stress were obtained.

Measurements were made also on the compression side of the fatigue specimens. No relaxation was observed (Fig. 15).

To explain the relaxation phenomena, the values obtained after fatigue testing are given in the Table 6, in comparison with results obtained on tension side. These results are compared with the tensile equivalent stress (von Mises criteria) calculated using the stresses values (fatigue stresses and residual stresses) observed at the maximum state of the fatigue loading.

These results seem to confirm the following hypotheses:

- There is a relaxation of the residual stress when the cyclic yield stress (Rev) of the metal is reached (Rev = 760 MPa for 42CD4 steel at 40 RC hardness. Rev is measured using cyclic stress strain curve plotting during constant strain amplitude fatigue test.)
- The value of the stabilized residual stresses, σRS is related approximately to the cyclic

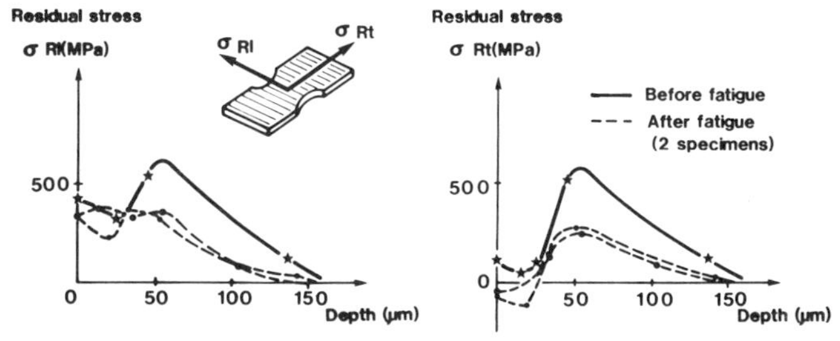

FIG. 12—*Residual stresses before and after fatigue testing in the case of hard grinding (condition DT) (results after fatigue testing from two test specimens).*

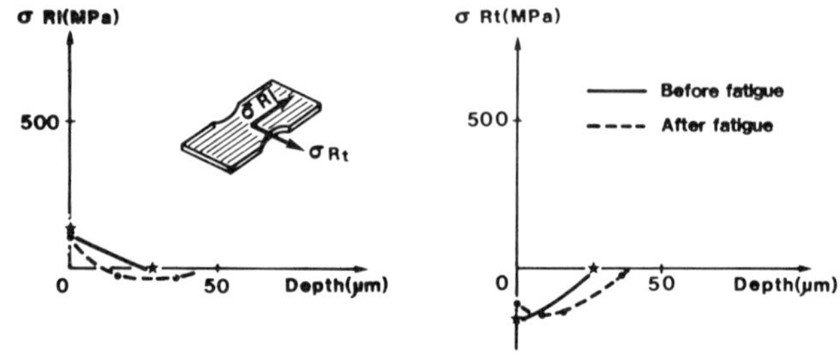

FIG. 13—*Residual stresses before and after fatigue testing in the case of soft grinding (condition AL).*

yield stress, Rev, of the metal and to the maximum stress, $\sigma_{f_{\max}}$, due to fatigue loading as seen in the following:

$$\sigma_{eq} \text{ Mises } (\sigma RS \cdot \sigma_{f_{\max}}) = \text{rev}$$

Obviously, these hypotheses are very rough. It is clear that the results will be different if the cyclic behavior of the materials is different (cyclic hardening or cyclic softening). Cyclic yield stress is insufficient to correctly predict the residual stress relaxation during fatigue loading. For this purpose the complete cyclic stress strain curve must be used. For example, to obtain a better prediction of residual stress relieving, it is possible to use a simplified method proposed by J. Zarka and J. Casier [*5*] from which a computer program using finite element analysis can be developed. Results obtained using this method are presented in Ref *6* for the case of grinding and shot peening.

The residual stress relaxation phenomenon is explained by means of a purely mechanical model (stressing beyond the cyclic yield stress). It should be kept in mind, however, that the residual stresses can be reduced also by means of loadings that are not always greater than macroscopic yield stress. This reduction can, for example, be obtained by vibrations, where the loading levels are well below the cyclic yield stress [*7*]. On the other hand, heat treatment can also be used [*15*] to reduce residual stresses (Fig. 16).

It can be seen that an explanation based purely on macroscopic behavior cannot by itself take into account all the phenomena involved in the reduction of residual stresses. Meanwhile, considering the well known definition of first order (macroscopic), second order (with the grain related stresses), and third order (with the crystal-related stresses) residual stresses

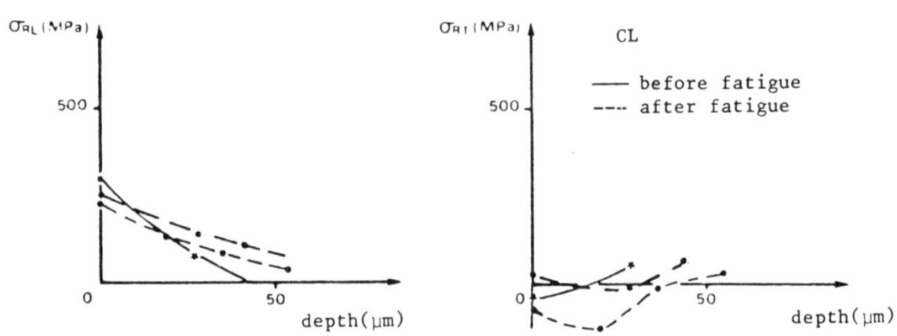

FIG. 14—*Residual stresses before and after fatigue testing in the case of medium grinding (condition CL) (results after fatigue testing from two test specimens).*

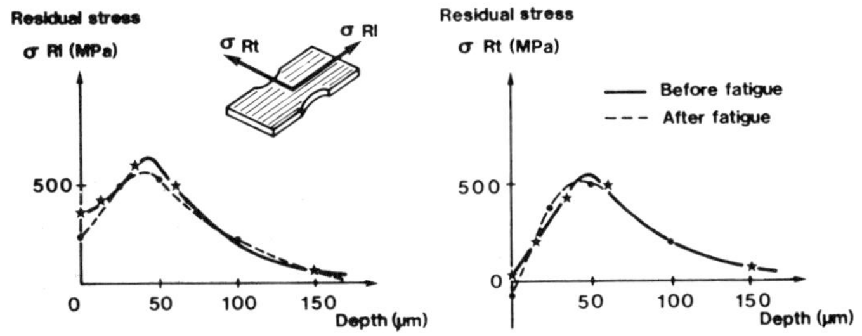

FIG. 15—*Residual stresses, before and after fatigue testing, on the compression side of test specimens (hard grinding condition DL).*

[*8*], it appears to be possible to explain the reduction in the residual stresses for loadings that are below the macroscopic yield stress. In the case of stress relief by vibration with a low macroscopic stress level, it is possible to imagine that, for those grains most adversely aligned in relation to the loading, the yield stress is exceeded locally and that the residual stresses within such grains are altered. This alteration will lead to a change in the macroscopic stresses (first order stresses), which are the sum of the second and third order residual stresses.

Calculated Prediction of the Fatigue Strength when Residual Stresses are Present

To compare and to validate various methods of calculation, values for the stabilized residual stresses measured after fatigue testing were introduced into various criteria as mean stresses [*16*]. The results obtained were then compared with a reference fatigue chart (Haigh or Goodman) when available. In the absence of such a chart, the validity of the criterion could only be investigated by examining the correlation between the experimental results. For the calculation, the maximum tension residual stresses measured after fatigue testing at the fatigue limit was introduced in the equations.

Haigh or Goodman Diagram

In the traditional method, using the residual stresses with Haigh or Goodman fatigue chart, only the residual stress parallel to the direction of fatigue stress is accounted for, and the transverse residual stress is totally neglected.

TABLE 6—*Calculation of the tensile equivalent stress (Mises criteria) at the maximum value of the fatigue loading.*

Type of Grinding and Measurement Face	Fatigue Stress, σf_{max}, MPa	Stabilized Residual Stress σR_l, MPa	Stabilized Residual Stress σR_t, MPa	Equivalent Stress σ_{eq} (Mises), MPa
Longitudinal grinding, extended face	+564	300	+320	757 relieving
Longitudinal grinding, compressed face	−564	+523	+501	522 no relieving
Transverse grinding, extended face	+520	+350	+285	698 relieving

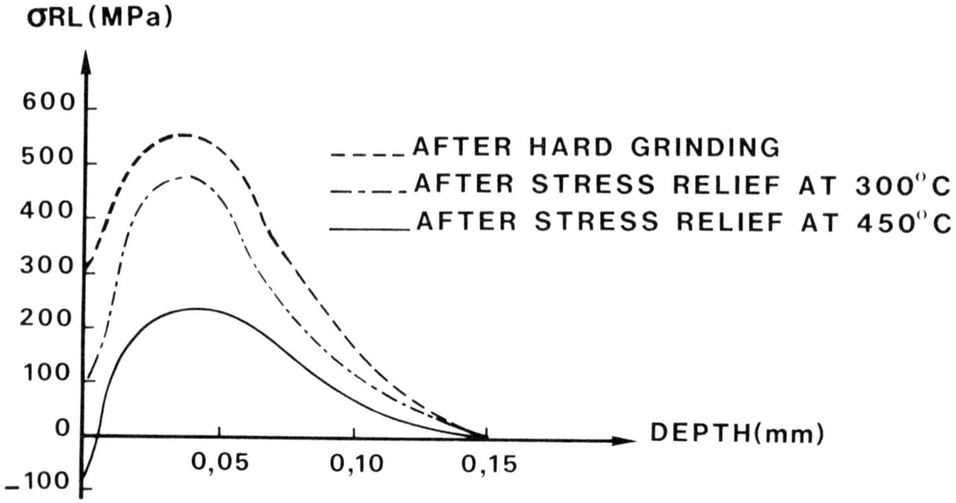

FIG. 16—*Reduction of the residual stresses σ_{RL} related to the stress relief temperature (1 h), for the hard grinding case.*

The calculation is performed by simply plotting on the σ_a (alternating stress)/σ_m (mean stress) fatigue chart the points obtained by experiments that are defined by the coordinates:

σ_a = amplitude of the fatigue stress at the fatigue limit, and

σ_m = mean stress of the fatigue stress, to which has been added the residual stress parallel to the fatigue stress.

The results obtained are plotted in Fig. 17. It is obvious that the method does not work correctly.

For this reason, this method cannot be generalized, and cannot be used unless it has been justified beforehand by means of experiments for the specific material, the manufacturing process, and the load.

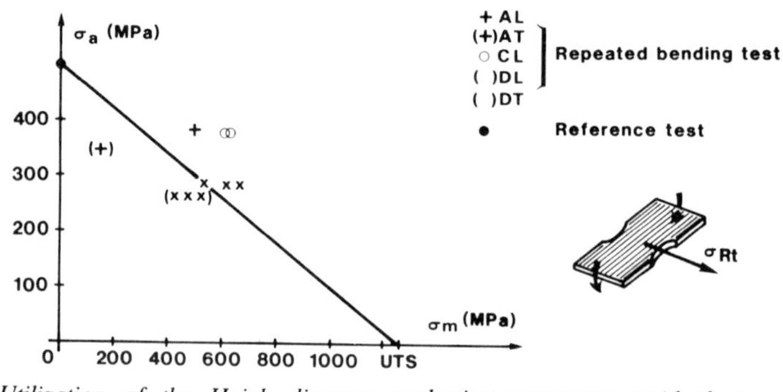

FIG. 17—*Utilization of the Haigh diagram neglecting transverse residual stress with respect to stress due to fatigue.*

Multiaxial Fatigue Criteria

From the complex state of stress, the von Mises criterion used for plasticity problems allows for the calculation of an equivalent tensile stress. Kiocecioglu [9] proposes for the fatigue problems the application of this calculation to the alternating stress and to the mean stress. In the case of fatigue cracking initiated at the surface, the state of stress is biaxial, and the equivalent stresses can be calculated from the relationships:

$$\sigma_{eq\ m} = (\sigma_{1m} + \sigma_{2m} - (\sigma_{1m} \cdot \sigma_{2m})^{1/2} \quad (1)$$

$$\sigma_{eq\ a} = (\sigma_{1a} + \sigma_{2a} - (\sigma_{1a} \cdot \sigma_{2a})^{1/2} \quad (2)$$

where

σ_a = alternating stress, and
σ_m = mean stress.

This equivalent tensile stress is related with the octahedral shear (τ_{oct}) stress following Eq (3):

$$\tau_{oct\ a} = \frac{\sqrt{2}}{3} \sigma_{eq\ a} \quad (3)$$

$$\tau_{oct\ m} = \frac{\sqrt{2}}{3} \sigma_{eq\ m} \quad (4)$$

The term $\sigma_{eq\ a}$ is calculated into Eq 2 using the amplitude of the principal stress, σ_1 and σ_2. The term σ_m is the mean stress due to the fatigue load and σ_{Rl}. The term σ_{Rt} represents the longitudinal and transverse residual stresses with respect to the orientation of the fatigue load. The two terms are introduced into Eq 1 through the definition of σ_{1m} and σ_{2m}, the two mean principal stresses, which, for the plane bending tests:

$$\sigma_{1m} = \sigma_m + \sigma_{Rl}$$

$$\sigma_{2m} = \sigma_{Rt} + \nu\sigma_m$$

where

σ_m = mean stress due to fatigue load and σ_{Rl}, and
σ_{Rt} = longitudinal and transverse residual stresses due to orientation of the fatigue load.

For each experimental result considered, both $\sigma_{eq\ a}$ and $\sigma_{eq\ m}$ were plotted on the Haigh reference chart. The results in Fig. 18 show that the points which represent the results are not on the reference curve. Therefore, this extrapolation of the von Mises criterion to fatigue in the presence of mean or residual stresses does not appear to be justified.

Figure 19 presents results obtained with the criteria proposed by Sines [*10*], Crossland [*11*], Findley [*12*], Matake [*13*], and Dang Van [7]. These criteria are written as follows:

Sines criterion:

$$\tau_{oct\ a} + \alpha \cdot p_m = \beta \quad (5)$$

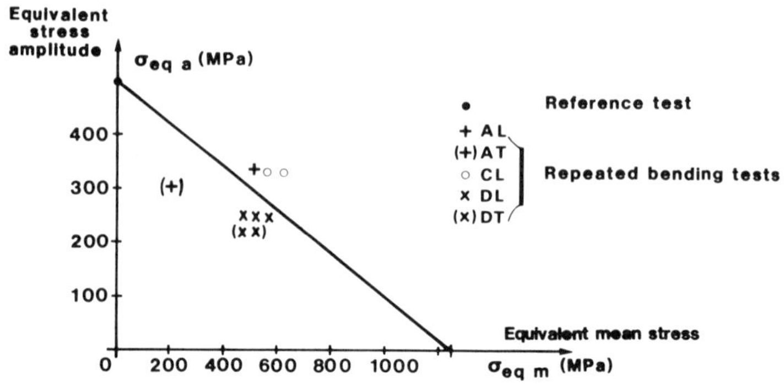

FIG. 18—*Experimental results interpreted using Kiocecioglu criterion on the reference diagram.*

Crossland criterion:

$$\tau_{oct\ a} + \alpha \cdot p_{max} = \beta \tag{6}$$

Findley-Matake criterion:

$$\tau_a + \alpha \cdot \sigma_n = \beta \tag{7}$$

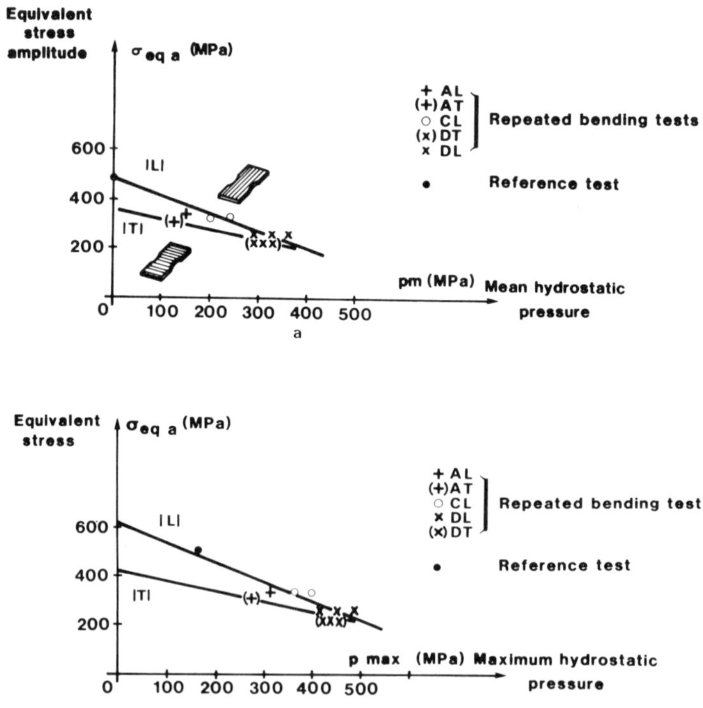

FIG. 19—*Experimental results interpreted using Sines (a), Crossland (b), Findley-Matake (c), and Dang Van (d) criteria. The reference test used here consists of a rotating bending fatigue test.*

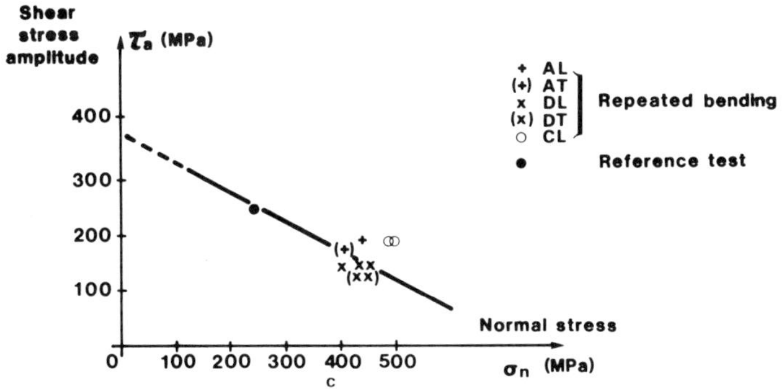

c

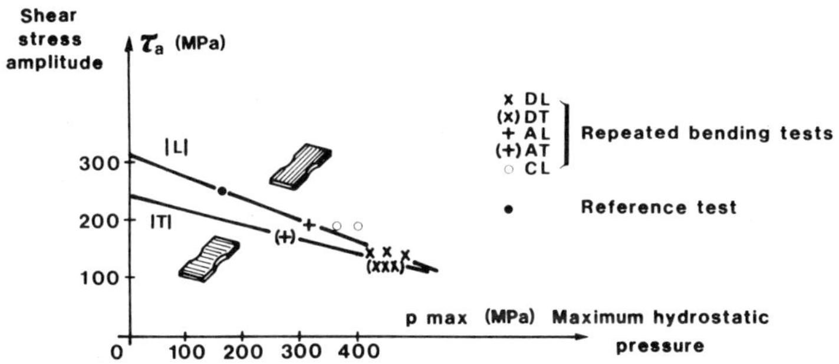

d

FIG. 19—*Continued.*

Dang Van criterion:

$$\tau_a + \alpha \cdot p_{max} = \beta \tag{8}$$

where

α, β = constants,
τ_a = maximum shear stress amplitude acting on the maximum shear plane,
$\tau_{oct\ a}$ = octahedral shear stress amplitude,
σ_n = normal stress perpendicular to the maximum shear plane,
p_m = mean value of the hydrostatic pressure during the fatigue cycle, and
p_{max} = maximum value of the hydrostatic pressure during the fatigue cycle.

Using the relationships in Eq 3, the Sines and Crossland criteria can be written in terms of $\sigma_{eq\ a}$ instead in terms of $\tau_{oct\ a}$ as presented on Figs. 19a and 19b. In Eqs 5 and 6, $\tau_{oct\ a}$ represents the shear stress amplitude that can be calculated using the Eqs 2 and 3 from the principal stress amplitude σ_{1a} and σ_{2a} (in biaxial stress state). The terms p_m and p_{max} correspond to the mean value and to the maximum value, respectively, of the hydrostatic pressure during the fatigue cycle. The hydrostatic pressure is the normal stress acting perpendicularly to the octahedral shear-plane and can be expressed in terms of the principal stresses (σ_1, σ_2, σ_3) using Eq 9:

$$p = (\sigma_1 + \sigma_2 + \sigma_3)/3 \tag{9}$$

In the case of the fatigue tests considered here the mean hydrostatic pressure is equal to one third of the sum of the mean stresses of the fatigue loading, including residual stresses as follows:

$$p_m = ((1 + \nu)\sigma_m + \sigma_{Rl} + \sigma_{Rt})/3 \quad (10)$$

and the maximum hydrostatic pressure can be written as follows:

$$p_{max} = [(1 + \nu)(\sigma_m + \sigma_a) + \sigma_{Rl} + \sigma_{Rt}]/3 \quad (11)$$

where

σ_m = mean stress of the fatigue loading,
σ_a = fatigue stress amplitude,
σ_{Rl} = residual stress parallel to the fatigue stress direction,
σ_{Rt} = residual stress perpendicular to the fatigue stress direction, and
ν = Poisson ratio of the material.

The plus sign used before the Poisson ratio in Eqs 10 and 11 is due to the plane strain state of the fatigue sample during the bending loading.

In Eqs 7 and 8 the term τ_a corresponds with the maximum shear stress amplitude acting on the maximum shear plane, and the term σ_n is the normal stress perpendicular to this plane. For the repeated bending fatigue tests presented here:

$$\tau_a = \sigma_a/2 \quad (12)$$

$$\sigma_n = (\sigma_a + \sigma_m + \sigma_{Rl})/2 \quad (13)$$

By comparing results presented on Fig. 19, the following conclusions can be drawn:

- The Sines, Crossland, and Dang Van criteria seem to agree well with the results obtained by experimentation. The amplitude of the octahedral shear stress and the hydrostatic pressures appear to be useful parameters for describing the behavior of a material subjected to multiaxial stresses.
- In Figs. 19a, 19b, and 19d two straight lines are obtained, one representing the intrinsic behavior for steel (longitudinal grinding with respect to the test specimen axis), and the other corresponding to the effect of surface finish (transverse grinding).
- The results obtained by application of Findley-Matake criterion are plotted in Fig. 19c, but these do not provide a clear correlation between the points obtained by experimentation and those corresponding to the linear equation proposed. Further, the linear extrapolation for alternating torsion ($\sigma_n = 0$) gives rise to values for the alternating torsion fatigue limit that are much too high. Finally, in the case of the results for grinding, this criterion does not provide for any separation of the effects of surface roughness.

Critical Layer Depth Criterion

It is also possible to interpret the results by means of the critical layer criterion which is an extension of the Dang Van criterion, and takes into account the stress gradient [*14*].

The principle of this criterion is simple: in order to take into account the physical phenomena acting during the initiation of fatigue cracks, especially the shearing of crystalline planes in a grain of metal, it should be more pertinent to consider the mean values of the amplitude of the shear stress, τ_a, and of the hydrostatic pressure, p_{max}. These mean values are calculated for a given thickness that is characteristic of the material and its structural state. This thickness corresponds to the elementary volume of the damaging process. The critical layer depth is defined as half the thickness of this characteristic volume. For quenched and tempered steels, the depth of this critical layer seems to be of the order of 50 μm [*14*]. This mean value of the stresses is then calculated using twice this thickness, that is, 100 μm.

If the mean value of the stabilized residual stresses over 100 μm is used as date for the calculation, the curves in Fig. 20 are obtained. As for Figs. 19a, 19b, and 19d, the experimental points representing the different machining conditions are broadly distributed along the straight lines of Eq 8.

As in the charts in Figs. 19a, 19b, and 19d, the residual stresses are taken into account in the calculation of the hydrostatic pressure. The upper lines represent the intrinsic fatigue strength of the steel, since in longitudinal grinding, the roughness has no influence (grinding marks parallel to the fatigue stress direction). The lower lines, on the other hand, show the influence of surface roughness.

From Figs. 17 to 20, the following statements can be made:

- When using a Goodman or Haigh diagram errors can be made because transverse residual stresses are neglected.
- Mises criterion cannot be used with loading that includes mean or residual stresses.
- The criteria studied that allow the residual stresses to be taken into account, through the hydrostatic pressure, give a good correlation of the experimental results and separate the respective influence of residual stresses and roughness.
- The comparison of the Findley-Matake and Dang Van criteria shows that the fatigue behavior is dependent upon the maximum hydrostatic pressure rather than the stress normal to the plane of maximum shear.
- The critical layer criterion seems to be usable for taking stress gradient into account.
- In order to validate any of the criteria studied, a greater number of experimental results concerning the grinding influence or the reference fatigue limit of the steel are required. The torsional fatigue limit of the steel would, for example, have allowed the lines drawn through the middle of the experimental points to be validated.

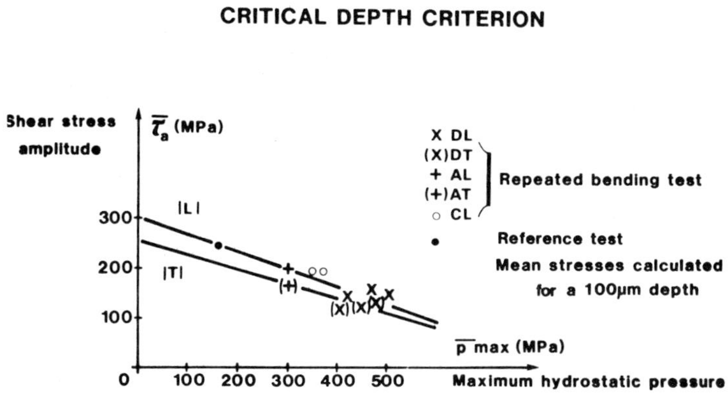

FIG. 20—($\overline{\tau}_a$, $\overline{p}_{max}$) *chart obtained with the mean values calculated for a thickness of 100 μm of the elementary volume.*

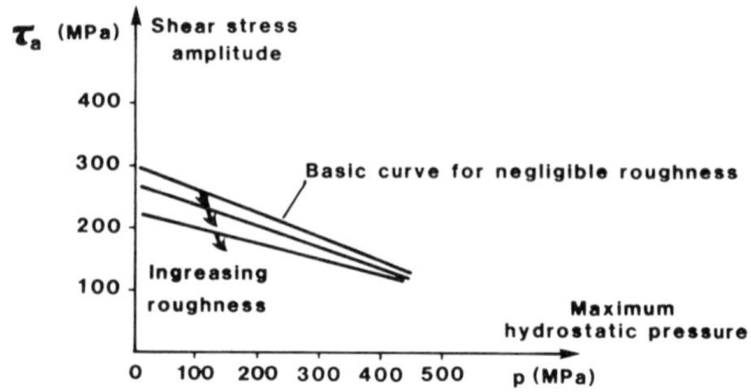

FIG. 21—*Plotting of the Dang Van chart for different values of roughness.*

Influence of the Surface Roughness

By using the Sines, Dang Van, or Crossland criteria, it has been possible to separate the respective influences of residual stresses and roughness. Residual stresses are taken into account in the hydrostatic pressure calculation as a mean stress effect. Roughness effect appears in the shear stress amplitude observed and is plotted in Fig. 21.

It should be noted, however, that with this approach it is difficult to take the influence of the surface roughness correctly into account for loadings as different as torsion and tension. It is known, in fact, that machining marks perpendicular to the axis of a test specimen do not influence in the same way the fatigue limit in tension and in torsion.

Conclusions

In this investigation of fatigue of a ground 42CD4 steel, a marked decrease in fatigue strength was seen when the severity of grinding increases.

X-ray diffraction was applied in tracking the relaxation of the residual stresses. When steel was subjected to a loading corresponding to its fatigue limit, a reduction of the order of 40 to 50 % was noted. This decrease seems to be linked to the cyclic yield stress of the metal being locally exceeded. To establish a model of this relaxation, it is necessary to account for both the cyclic behavior of the metal and microplasticity phenomena.

Taking into account the residual stresses, criteria such as the Dang Van or the Crossland, which make use of the hydrostatic pressure, allow the respective influences of the surface roughness and the residual stresses on the fatigue strength to be separated. The use of Dang Van or Crossland fatigue charts, plotted for different values of roughness, should help design engineers to better quantify the factors that affect the fatigue strength of a mechanical engineering part such as the following:

- Mechanical properties and microstructure of the material,
- Surface roughness, and
- Residual stresses.

References

[1] Brand, A. and Flavenot, J. F., *Recueil de Données Technologiques sur la Fatigue* (Fatigue Datas for Mechanical Engineering), CETIM, 1981.

[2] Flavenot, J. F. and Skalli, N., "Influence of Grinding Conditions on the Residual Stresses Introduced in a 42CD4 Grade Steel," *CETIM Information,* No. 71, 1981.

[3] *La Pratique des Essais de Fatigue* (*"Practice of Fatigue Testing"*), PYC, Paris, 1982.

[4] Dölle, H. and Cohen, J. B., "Residual Stress in Ground Steels," *Metallurgical Transactions A,* Vol. 11 A, January 1980, pp. 159–164.

[5] Zarka, J. and Casier, J., "Elastic Plastic Response of a Structure to Cyclic Loading: Practical Rules," *Mechanics Today,* Vol. 6, 1979, pp. 93–198.

[6] Lu, J. and Flavenot, J. F., "Use of a Finite Element Method for the Prediction of Residual Stress Relaxation During Fatigue," *Mechanical Relaxation of Residual Stresses, ASTM STP 993,* ASTM, Philadelphia, 1988.

[7] Dang Van, K., Cailletaud, Flavenot, J. F., Le Douaron, and Lieurade, H. P., "Multiaxial Long Life Fatigue Criteria," *Proceedings of the Journées Internationales de Printemps,* Société Française de Metallurgie, Paris, 22–23 May 1984, pp. 301–337.

[8] Hauk, V., Eigenspannungen, Ihre Bendentung für Wissenschaft und Technick" ("Residual Stresses, their Understanding for Science and Technology"), *Eigenspannungen,* Ed. by Deutsche Gesellschaft für Metallkunde (Germany), Vol. 1, 1983, pp. 9–49.

[9] Kiocecioglu, D., Stultz, J. D., and Nofl Jr., C. F., "Fatigue Reliability with Notch Effects for AISI 4130 and 1038 Steel," *ASME Transactions, Journal of Engineering for Industry,* February 1975, pp. 359–370.

[10] Sines, G., "Fatigue Criteria Under Combined Stresses or Strains," *Transactions ASME, Journal of Engineering Materials and Technology,* Vol. 103, April 1981, pp. 82–90.

[11] Crossland, B., "Effect of Large Hydrostatic Pressures on the Torsional Fatigue Strength of an Alloy Steel," *Proceedings of the International Conference of Fatigue of Metals,* Institution of Mechanical Engineers, London, 1956, pp. 138–149.

[12] Findley, W. N., "A Theory of the Effect of Mean Stress on Fatigue of Metals Under Combined Torsion and Axial Load or Bending," *Transactions ASME, Journal of Engineering for Industry,* Vol. 81, 1959, p. 301.

[13] Matake, T., "An Explanation of Fatigue Limit Under Combined Stress," *Bulletin of the JSME,* Vol. 20, No. 141, March 1977, pp. 257–263.

[14] Flavenot, J. F. and Skalli, N., "A Critical Depth Criterion for the Evaluation of Long Life Fatigue Strength Under Multiaxial Loading and a Stress Gradient," *Proceedings of the Fifth European Conference on Fracture,* ECF5, Lisbon, 17–21 September 1984, p. 335.

[15] Bouhelier, C., "Heat Treatment for Pressure Vessels," *Proceedings of the Fourth Conference on Pressure Vessels,* Association Française des Ingénieurs en Appareils à Pressions (AFIAP), Paris, Vol. 3, October 1983, pp. 219–254.

[16] Skalli, N. and Flavenot, J. F., "Taking into account Residual Stress in Fatigue Design," *CETIM Information,* No. 90, May 1985, pp. 35–47.

Summary

As a point of departure in summarizing the papers in this volume, consider a straight bar of metal that contains a distribution of residual stresses resulting from the fabrication or the heat treatment of the bar. Generally, this distribution will consist of a relatively thin skin that is in longitudinal tension or compression, with the internal material or the core carrying residual stresses of the opposite sign. These stresses are a manifestation of the small elastic deformations that the different elements of the bar must experience in order for them to remain compatible, that is, in order for them to stay fitted together. If the bar is now pulled in tension with a force exceeding the yield load then much of the elastic deformation will be relieved by plastic flow and, after unloading, the residual stresses will have been relaxed. If the stress-strain curve for the bar material is flat and horizontal in the plastic range then the residual stresses will have been removed entirely. However, even if the material exhibits some strain hardening in the plastic range, considerable relaxation of the residual stresses is achieved by this mechanical treatment.

Conceptually, this is the simplest approach to the mechanical relaxation of residual stresses, and it is this approach which *Nickola* verified so elegantly in the first paper in this volume. Using a carefully conceived and executed series of experiments on an experimental aluminum alloy, he showed first how a fabrication process—in this case cold rolling—can change the residual stress distribution in a plate rather drastically. Then he demonstrated the efficacy of cold stretching in reducing these stresses. In so doing, *Nickola* employed two different methods of residual stress measurement; the standardized hole-drilling strain-gage method and a version of the layer-removal method. The excellent agreement between the results obtained by the two methods provides a supplementary benefit of *Nickola's* work; it suggests that the layer-removal technique for residual stress measurement that he used may well be a promising candidate for standardization.

All of the papers in this volume were based upon research that had been initiated long before the call for papers for the symposium was issued. Therefore, although all of the papers directly address the mechanical relaxation of residual stresses, they also provide interesting and valuable results on related topics. *Nickola's* excellent results with the layer removal method constitute an example of this.

Altschuler, Kaatz, and Cina also showed how plastic deformation in tension could be used very effectively to relax stresses. (Their work involved 7075 aluminum alloy plate in the as-quenched condition, and X-ray diffraction measurements of the residual stresses.) Recognizing, however, that tensile loading is not practical for objects of all shapes—various forgings, for example—they also studied stress relaxation under compressive loading. They found that plastic deformation in compression was effective in relaxing residual quenching stresses, but not quite as effective as tensile deformation. In both cases, maximum (although incomplete) relaxation of residual stresses was achieved after less than 1.5% deformation. The test data suggest that substantial deformation beyond this point may have reversed the process and begun to increase the residual stress levels slightly.

We might submit that the superior effectiveness of tensile deformation for the relaxation of residual stresses in 7075 aluminum alloy may stem from the lesser slope of the plastic stress strain curve for the material in tension as compared with compression. In other words, the development of strain hardening in this material is slightly more gradual in tension than in compression.

Altschuler et al. include an interesting discussion in their paper of various aspects of cold working as it might be used to reduce residual stresses in practical situations, for example, by compressing an irregularly shaped forging in a die. Related results obtained by earlier investigators are cited extensively.

Our understanding of residual stress relaxation by cold working processes is further enhanced by *Leggatt and Davey,* whose work involved welded alloy steel panels. Their research investigated the situation where the mechanical forces that are applied are not of magnitude sufficient to stress the entire object into the plastic range. They showed that wherever the residual stress is locally of sufficient magnitude such that when added to the applied stress the yield strength is exceeded, then some relaxation of the residual stress will occur. In general, we might expect the amount of relaxation to be simply the amount by which the yield strength is locally exceeded. Since the residual stresses in welded joints are usually about equal to the yield strength, the relaxation in *Leggatt and Davey's* specimens was approximately equal to the applied stress.

They showed, further, that, depending upon the geometry of the object, stress concentration factors may alter the applied stresses from the nominal values corresponding to the applied mechanical forces. The welded specimens that *Leggatt and Davey* used were fabricated with intentional misalignments and distortions which provided effective stress concentration factors. Their work demonstrated that if the object contains reentrant corners or sections in bending, the effective stress concentration factors could be negative.

Leggatt and Davey's residual stress measurements were made by the hole-drilling method. The research they reported was part of a larger project whose aim was to determine allowable defect sizes in welded spherical ammonia storage vessels that were subject to proof testing before entering service. On the basis of the research reported here they suggested that the total stresses on the welds in service (the sum of applied and residual stresses) may be assumed to be equal to the yield strength of the region in which the defect is located.

Ohol, Nagendra Kumar, and Noras were also concerned with fabricated steel structures containing residual stresses resulting from welding. Their motivation for stress relaxation was to achieve dimensional stability. Their selection of a method for achieving stress relaxation was based upon: (1) the fact that their structures are geometrically complex; and (2) research in the literature according to which relatively small relaxations of residual stresses are usually sufficient to provide dimensional stability. (We might suppose that for dimensional stability residual stresses need only be reduced to levels at which creep is negligible at the operating temperatures; although, if further machining of the structure must be done after partial stress relaxation, some distortion could result.)

Ohol et al. chose controlled vibration as the most practical and economical means for achieving mechanical relaxation of the residual stresses in their complex structures. The principle of this method is that large displacements and strains can be produced as a result of a structure resonating under the influence of a time-varying force of relatively small amplitude. In selecting this method, *Ohol et al.* used the same reasoning as *Leggatt and*

Davey, that is, the applied stress must be of a magnitude such that the algebraic sum of the applied and residual stresses exceeds the yield strength of the material, causing plastic deformation.

Since the strain amplitude during resonance is not uniform over a structure, differing levels of residual stresses remain after the mechanical treatment. *Ohol et al.* surmise that the relaxation would have been greatest at regions of high stress concentration and at surfaces where the highest strain amplitudes were imposed due to vibration of the structure in bending modes.

Vibratory stress amplitudes about half of the yield strength were attained, and were imposed for thirty seconds at each of two natural frequencies. The result was relaxation of 30 to 57 % of the residual stress magnitudes and the effect was good stability of dimensions.

Ohol et al. used the hole-drilling method to evaluate the stress relaxations that they achieved. Although this method is reliable, it is expensive and tedious. Therefore, they propose that in production situations vibratory strain amplitude could serve as a measurable quantity to determine whether adequate stress relaxation will occur at the location of interest.

Bouhelier, Barbarin, Deville, and Miege also studied the mechanical relaxation of residual stresses in welded steel components. Like *Ohol et al.,* they chose mechanical vibration as the means to effect the relaxation. Due to an error, however, they conducted their first vibration treatment at a frequency only two-thirds of the structure's first natural frequency. After vibrating the structure at this frequency for 15 min they measured 45 to 100 % relaxations of residual tensile stresses, and 0 to 45 % relaxations of residual compressive stresses despite the fact that the magnitude of the vibratory stresses did not exceed 1.5 MPa (220 lbf/in.^{2}). Having identified their error, *Barbarin et al.* repeated their vibration treatment of the structure, this time at its first natural frequency. Although this treatment generated vibratory stresses more than 100 times greater than those obtained in the first test, no further changes in the residual stresses were detected.

The dominant position in the literature regarding vibratory stress relief has been that it should be carried out at resonant frequencies in order to be effective. There have been a few reports suggesting that sub-resonant vibration may be just as effective and, perhaps, more so. To this editor's knowledge, however, this paper by *Barbarin et al.* is the first that documents this phenomenon with actual measurements of the stress relaxation. (They used X-ray diffraction techniques to measure residual stresses before and after the vibration treatments.)

Now it is abundantly clear in the case of this relaxation by sub-resonant vibration that the vibrations did not cause the yield strengths of the material to be exceeded, even locally. So the mechanism of the relaxation was something other than gross yielding. *Bouhelier et al.* suggest that the mechanism was, perhaps, movements and reorganization of anomalies at the atomic level, for example, dislodging of dislocations, movement of interstitial atoms, internal friction, etc.

The *Bouhelier et al.* paper raises still another question. They found that tensile residual stresses were more easily relaxed than compressive residual stresses. Why? We might conjecture that this observation merely reflects the fact that residual stress measurements could only be made at a limited number of discrete locations. However, every body contains both tensile and compressive residual stresses. For reasons of equilibrium it is clear that reductions in tensile residual stresses in a body must be accompanied by comparable reductions in the compressive residual stresses.

Moving on from intentionally applied dynamic stresses (vibratory stress relief) to unintentional ones (fatigue), *Lu, Flavenot, and Turbat* conducted a combined analytical and experimental program to study the means by which residual stresses that had been introduced by shot peening and by grinding are relaxed. Using cyclically softening alloy steels (that is, steels whose yield strengths decrease with continued applications of stress cycles into the plastic range), they showed that the relaxation of the residual stresses under fatigue loading increased with cyclic stress amplitude and with the number of cycles applied. For a given cyclic stress amplitude, the relaxation of compressive residual stresses increased as the minimum stress was made more negative. These changes in the residual stresses were measured by X-ray diffraction techniques. It was hypothesized that the principle of stress relaxation that was operative here is that, instead of introducing stresses which cause the yield strength to be exceeded, the yield strength is reduced by cyclic stressing to a level less than the sum of the applied plus residual stresses.

Lu et al. then calculated the effects of fatigue loading on the relaxation of residual stresses for comparison with their experimental measurements. The calculations were based on a model developed elsewhere, and were carried out with the aid of a finite element computer program also developed elsewhere. As input data, the approach requires cyclic stress-strain curves for the material. In very simplistic terms, it seems that the relaxation of the residual stresses in the surface layer is numerically estimated; then the effect of the altered stress distribution in the surface layer on the stresses in the first subsurface layer is determined; then the effects on the next subsurface layer; etc. Although not explicitly stated, it appears that an iterative procedure must be used in order to eventually arrive at stress distributions which satisfy the conditions of equilibrium, compatibility and the material stress-strain characteristics. According to the authors, the solutions converge after only a few cycles.

The calculated stress relaxations show qualitative agreement with the measured stress relaxations. *Lu et al.* attribute the differences to the fact that they used cyclic stress-strain curves for virgin material whereas the surfaces of their specimens had been shot peened or ground. The degree of agreement achieved offers promise of an approach toward something more than a superficial understanding of the relaxation of residual stresses under fatigue conditions.

In the final paper, *Flavenot and Skalli* studied the effects of grinding on the fatigue behavior of a steel. Under certain grinding conditions some surface roughness as well as residual stresses are produced in the material. By measuring the relaxation of these stresses by X-ray diffraction techniques as the fatigue process proceeded, the authors were able to estimate the separate effects of the surface roughness and the residual stresses on the fatigue behavior. The fatigue tests were carried out in bending. It was noted that the tensile residual stresses were not relaxed on the specimen surface which was cyclically stressed in compression. This observation is certainly consistent with the mechanism of residual stress relaxation which depends upon the yield strength being exceeded. Despite this, *Flavenot and Skalli* caution the reader that residual stresses can also be relaxed by mechanisms which do not exceed the yield strength, which is a reference to vibratory treatments with low dynamic stress levels.

It is largely coincidental but fortunate, nevertheless, that the seven contributed papers that comprise this volume cover the subject of mechanical stress relaxation so well. They address all three of the principal mechanisms: cold working or overstressing, vibratory stress relief, and fatigue cycling. Furthermore, all of the papers are based upon actual measurements of relaxation, in contrast to many previous studies which were inferential and qualitative. Thus, it is likely that this volume will serve as a reliable foundation for those needing an introduction to the mechanical relaxation of residual stresses, as well as a current review for those seeking to research or otherwise advance the state of knowledge in this technologically important field.

To be sure, questions remain for further research. The mechanism of stress relief under sub-resonant vibration and apparent differences in the relaxation of tensile and compressive residual stresses need clarification. However, the basic concepts of the relaxation processes now appear to be reasonably well understood. This will enable these processes to be exploited more fully where they will be beneficial, and avoided or accounted for where they would be detrimental. It is, perhaps, not too soon to begin efforts to standardize selected aspects of mechanical relaxation processes for residual stresses.

Leonard Mordfin
United States Department of Commerce
National Bureau of Standards
Gaithersburg, Maryland; symposium chairman and editor

Author and Discussor Index

Subject Index

S

T–X